Taking Advantage of Emergence

Taking Advantage of Emergence

Productively Innovating in Complex Innovation Systems

Deborah Dougherty

OXFORD
UNIVERSITY PRESS

OXFORD
UNIVERSITY PRESS

Great Clarendon Street, Oxford, OX2 6DP,
United Kingdom

Oxford University Press is a department of the University of Oxford.
It furthers the University's objective of excellence in research, scholarship,
and education by publishing worldwide. Oxford is a registered trade mark of
Oxford University Press in the UK and in certain other countries

First Edition published in 2016
Impression: 1

Published in the United States of America by Oxford University Press
198 Madison Avenue, New York, NY 10016, United States of America

British Library Cataloguing in Publication Data
Data available

Library of Congress Control Number: 2015944566

ISBN 978–0–19–872529–9

Printed in Great Britain by
Clays Ltd, St Ives plc

Contents

1

It Takes an Infrastructure to Take Advantage of Emergence

> Emerge: to rise from or as from a surrounding fluid; to come forth into view; become visible; to develop or evolve as something new, improved, etc.
>
> (*Webster's New World Dictionary*, 4th edn)

> The system is dynamic, the whole is greater than the sum of the parts, and solutions can't be imposed; rather they arise from the circumstances. This is frequently referred to as *emergence*.
>
> (Snowden and Boone 2007: 7)

> Complexity theory spotlights emergence as its central phenomenon, helping to explain how system-level order spontaneously arises from the action and repeated interaction of lower level system components without intervention by a central controller.... Because order in self-organizing systems relies not on the imposition of an overall plan by a central authority, but on the action of interdependent agents purposefully pursuing individual plans based on local knowledge and continuously adapting to feedback about the actions of others, it is said to emerge spontaneously.
>
> (Hayek 1988; cited by Chiles et al. 2004: 502)

Many of society's most pressing problems are complex innovation systems. Health care, alternate energy, climate management, and myriad social problems from poverty to economic revitalization rely on ongoing innovation to continually create projects, processes, and strategies that address these critical challenges. Innovation needs the creation, combination, and recombination of knowledge about problems and possible ways to resolve them. But in complex systems, knowledge is fragmented and scattered, because we do not know the relationships between cause and effect, what particular elements may be involved, or how these elements might interact. Knowledge for innovation in complex systems also emerges unpredictably. The only way that we can grapple with these pressing societal problems is to take advantage of emergence: use it to spot minor perturbations that may escalate into significant problems or solutions, and to configure fragmented information bits into innovative solutions for significant problems.

This book is titled *Taking Advantage of Emergence* to highlight the enormous potential in complex innovation systems for learning and creating value, even though available information is fragmented, partial, and widely dispersed. Understanding how to take advantage of emergence is an important step in our ability to leverage the explosion of sciences and technologies into better resolutions for social and economic challenges. More than one hundred years ago, major transformations in what Nelson (2005) and Pisano (2010) call social technologies led to the rise of modern industrial society, because these social technologies enabled people to make use of steam power, steel making, mass production, and other physical technologies. These authors argue that making use of the possibilities from science-based complex innovation systems again requires major transformations in social technologies in the early twenty-first century. The collective capability for taking advantage of emergence is one such transformation in our social technologies for addressing big problems.

I use my research on developing new drugs—a complex innovation system—to create a detailed and systematic approach through which scientists, managers, and other knowledge professionals in complex systems can continually generate knowledge for innovation, and collectively figure out how to use that knowledge productively. Drug discovery informs and illustrates this new framework and I rely primarily on examples in this domain. But I hope that people working on other complex innovation systems can think about applying these ideas to their domains.

Nearly every field in the social, behavioural, and economic sciences can contribute to the development of taking advantage of emergence. However, I draw primarily on insights from the fields of managing innovation and of science, technology, and innovation, and on Karl Weick and others who develop ideas for collective learning. Together, these sets of ideas provide a useful foundation for taking advantage of emergence, because they combine the details of innovating with the social processes of collective learning. Other fields would definitely add important extensions, so I encourage those scholars to build on and reframe the ideas presented here. Innovation, by definition, refers to creating, developing, and implementing new products, new processes, and new strategies. Innovation goes much beyond simply coming up with novel ideas. Innovation creates, combines, and recombines novel ideas into actual, material things or programmes that function in the real world. Innovation management and science and technology highlight recent transformations in social technologies for managing and applying knowledge for innovation. These transformations serve as stepping stones towards the qualitatively new transformations in social technologies required for complex innovation systems.

Because complex new products like new drug therapies emerge from many elements with unknown interactions, innovators must discover both the specific elements that comprise a new product and the novel configuration of interdependencies among these elements that make the product functional in the real world. My framework for taking advantage of emergence builds on abductive reasoning, or the logic of discovery (Nesher 2001). Abductive reasoning provides a way to create, combine, and recombine knowledge into viable new products, processes, and strategies despite, or indeed because of, the inherent complexity. Taking advantage of emergence is a process of discovery. Abduction, first articulated by Charles Peirce and other pragmatic philosophers in the 1800s, is the deliberate reasoning that leads to scientific discoveries (Nesher 2001). According to Peirce, abduction is the best answer we have to problems of discovery, since abduction alone among the forms of reasoning originates possible explanations and introduces new ideas. Locke et al. (2008: 907) quote Peirce to explain: '[d]eduction proves that something *must* be; induction shows that something *actually is* operative; abduction merely suggests that something *may be*' (emphasis in original). Abduction 'is the process of reasoning in which explanatory hypotheses are formed and evaluated' (Magnani 2001: 18).

Paavola et al. (2006) explain how Darwin relied on abductive reasoning to develop his theory of evolution. Darwin, they say, consciously built a network of trusted fellow naturalists, animal breeders, and missionaries around the world to whom he could address questions. He joined breeders' clubs, established personal connections with pigeon raisers, and contributed to horticultural journals to solicit information.

> What is remarkable about this enterprise is not only the skill with which he organized and carried out the in-house anatomical dissections, but also the impressive managerial talents he exhibited in creating the scientific network that supported these activities. Darwin was not a mathematical whiz kid. His genius was very much in his ability to locate big questions, and in the strategic eye he had for soliciting information needed to answer these questions. (Paavola et al. 2006: 141–2).

Darwin took advantage of emergence by discovering possible theories to explain natural history, developing great questions from those theories that would illuminate these possibilities, drawing on dispersed knowledge from fellow naturalists and others, and continually refining and reframing his theories.

Paavola et al. (2006: 142) also quote Ernst Mayr on Darwin:

> [Darwin's] procedure does not fit well into the classical prescriptions of the philosophy of science, because it consists of continually going back and forth between making observations, posing questions, establishing hypotheses or models, and testing them by making further observations, and so forth.

This describes the whole process of abductive reasoning.

As the Darwin story suggests, taking advantage of emergence occurs in an infrastructure of physical technologies, artefacts, and testing regimes, and social technologies such as categories of knowledge, governance structures, and networks of relationships, skills, and experiences. This book focuses on the infrastructure necessary for complex innovation systems, and specifically on the social technologies that structure this infrastructure. I elaborate three sets of social technologies that enable innovators in complex innovation systems to take advantage of emergence and collectively discover viable new products. These social technologies encompass: (1) a new division of labour that highlights distinct subsystems of discovery, each with their own problem that must be addressed to take advantage of emergence; (2) abductive learning routines that animate discovery processes within and across the distinct subsystems; and (3) organizing through heedful interrelating that organizes the infrastructure and keeps knowing grounded in doing.

I draw on my research with several colleagues in pharmaceutical drug discovery to delve into taking advantage of emergence in complex innovation systems. We have published five journal articles so far on one of the subsystems or ideas (Dougherty 2007; Dougherty and Dunne 2011, 2012; Dougherty et al. 2013a; Dunne and Dougherty 2016), have more under review (Dunne 2015; Su and Dougherty 2015), and several more in the works. We have also published two book chapters (Dougherty et al. 2013b; Dougherty 2015). Each of these pieces of scholarship delves into one of the many facets of this complex innovation system. In this book, I attempt to synthesize all these ideas into a coherent framework.

Drug discovery is a complex innovation system that has been studied a great deal, so it provides a good empirical context to explore challenges and develop possible solutions. To innovate, drug discovery scientists seek to discover heretofore unknown patterns of interactions among molecular compounds, a disease, and the human body. However, as I will detail in this book, despite enormous progress, biomedical science cannot explain the causes of many medical conditions such as various forms of cancer, depression, autism, rheumatoid arthritis, or Alzheimer's disease. Innovators have great difficulty figuring out what proteins and cell pathways are implicated in the disease process, how to modify those specific pathways, and how to build molecules that can be absorbed into the body and circulated through the blood stream to the particular disease site without also interacting with other biological systems to cause unwanted side effects. One indicator of the complexity is that 50 per cent of studies on 'targets' published in academic journals cannot be replicated (Frye et al. 2011; Prinz et al. 2011). Targets refer to the proteins that are part of a pathway in a cell or on its surface that generate disease processes. Another indicator is that much heralded major break-throughs such as the map of the human genome have not yet led to any new cures (Cohen 2011). People had assumed that the human genome map would lead to cures because a limited number of common genetic variants underlie diseases. However, now it seems that even common heritable diseases are linked to many genetic variants, each relatively rare (Singer 2009).

Not surprisingly, complexity troubles pharmaceutical innovation. Despite enormous public and private investment in biomedical sciences, and despite the claims from organization and strategy scholars about the positive effects of new biotechnologies, the number of new drugs produced per $billion spent continues to decline (Scannell et al. 2012). The translation of basic biomedical research into safe and effective drugs remains slow, expensive, and failure prone; the average product cycle time is thirteen years, the failure rate still exceeds 95 per cent, and the cost per successful drug exceeds $1 billion after adjusting for all the failures (Collins 2011). But society values health care so we pump considerable public and private investment into drug discovery. There are many different discussions about the promises and problems in this industry that I will draw on to develop my framework. Other complex innovation processes such as improving education, revitalizing damaged ecologies, addressing poverty, or remediating inner cities receive less funding and less ongoing attention. Indirectly, these systems provide some ideas about what happens when aspects of the complex innovation system are ignored.

It may seem odd to focus on social technologies, since much of the popular and academic media focus on breakthroughs in physical technologies, and on the heroic accomplishments of individual scientists and entrepreneurs. However, according to Chandler (1977), the technological and scientific innovation that underpinned industrial society evolved interdependently with organizational and managerial innovation. New forms of business organization and new institutional arrangements play a critical role in facilitating technological advance and the diffusion of innovations. For example, advances in physical technologies such as steam power, steel making, mechanical engineering, and so on made railroads and mass production technically feasible. But a host of novel social technologies made these technological advances economically feasible. Among these social technologies are administrative hierarchies, professional managers, business schools to train those managers and engineering schools to train the technologists, formalized capital budgeting systems, and governance systems that separate ownership and management.

According to Nelson (2005: 208):

> Today, some of our most difficult problems involve developing the social technologies needed to make new physical technologies effective. Arguably the lion's share of the strains in our health care systems are the result of advances in physical and medicinal technologies that societies have not yet learned how to manage or pay for.

I suggest one way that allows society to make these new physical technologies effective.

Significant advances have been made in the pharmaceutical innovation system's physical technologies, including biotechnology (e.g. monoclonal antibodies, recombinant DNA), genomics, bioinformatics, proteomics, imaging technologies, high throughput screening, and many others. Both industry experts (e.g. Ng 2004; Collins 2011) and organization scholars (e.g. Arora and Gambardella 1994; Henderson et al. 1999) emphasize these physical technologies. But others who study pharmaceuticals argue that new social technologies for using all these physical technologies are yet to be developed (Barry 2005; Nelson 2005; Pisano 2006; Christensen et al. 2009; Dougherty and Dunne 2012). According to Pisano (2006), adding physical technologies stretches the already vast search landscape of drug discovery so new social means to integrate the technologies are necessary. West and Nightingale (2009) argue that many of the remaining bottlenecks in pharmaceuticals need social rather than technological solutions.

I will weave these diverse suggestions for transforming social technologies for this complex innovation system together with the large body of work on innovation and science and technology. The result is a new framework for taking advantage of emergence.

This book is intended to be academic in that I build on theory, synthesize a variety of conceptual thrusts into a coherent framework, and apply this synthesis of ideas to a critical social challenge. My intent is to understand this social challenge and develop ways to address it more effectively. However, I deviate from the stereotype of 'academic' in two ways that I hope make the book more palatable to non-academic scientists, managers, policy experts, and other knowledge professionals who wish to reflect on innovating better in complex innovation systems. One deviation is to avoid the academic tendency to argue endlessly over definitions. Instead, I select particular definitions that ring true to me, explain why, try to articulate the ideas clearly so anyone can understand what I mean, and then apply them. I think that understanding how ideas can come together and play out in practice enables social scientists to learn more about complex phenomena than do abstracted debates over possible meanings. Readers are warned that academics vigorously contest the ideas in this book: complexity, emergence, infrastructures, systems, routines, learning, and abduction are all defined by different people in different ways. I do not attempt to summarize all relevant literature, but do suggest references.

The second deviation is to edge towards the normative, and suggest what people should do—something that some academics declare we never should do. I am inspired by Dick Boland who, in his division keynote address at the Academy of Management meeting in 2013, emphasized the importance of taking a pragmatic view in our research. He said that we should become designers of better worlds, not simply observers of inadequate ones. In a similar way, Anna Grandori (2010) talks about trying to understand the best patterns of thinking that can be found in scientific work rather than summarizing what people on average do. I only 'edge' over to prescriptive advice, however, because complex innovation systems are complex, and there are no simple ideas or easy steps, and nothing can be determined or guaranteed. Those who want quick answers must look elsewhere, because my framework ends up with the idea that taking advantage of emergence for innovation is an ongoing process.

In the rest of this chapter, I describe and illustrate the concept of infrastructure. Then I outline three sets of social technologies that structure the infrastructure so that people can take advantage of emergence rather than avoid it or seek to eliminate it. Chapter 2 elaborates on the foundational role of abductive learning routines in complex innovation systems. Chapters 3, 4, 5, and 6 explain how these learning routines are specifically enacted in each of the four subsystems of the complex innovation system. Chapter 7 pulls the ideas together again, and suggests some next steps in developing the capability for taking advantage of emergence.

It Takes an Infrastructure

Taking advantage of emergence occurs across an infrastructure of physical and social technologies, not in individual firms, institutions, or industries. The infrastructure is the subject of study and the unit of analysis for taking advantage of emergence. People use many different terms for this 'big context' within which innovation takes place, including regime, field, ecosystem, community of populations, or simply 'the market'. I use the definition of 'infrastructure' as developed by scholars of science, technology, and innovation (Nightingale 2004) and economic growth (Nelson 2005) to define this big context. The science, technology, and innovation understanding of infrastructure embraces the emergence and co-evolution of knowledge and innovation, encompasses public and private enterprises and institutions, and highlights the human orientations and activities that the social technologies need to enable.

Nightingale (2004) summarizes a large literature on science, technology, and innovation that supports the need for a discovery style of research that seeks to understand, not just to predict (Pavitt 1999). Innovation arises from the convergence of fragmented information that researchers generate as they work in different institutions. An effective infrastructure enables people to interpret locally generated experimental knowledge in a cumulative manner by creating conditions for comparability between experimental conditions. The discovery style of research makes it possible for scientists to tinker with experimental conditions to identify and fine-tune more promising alternatives (Pavitt 1987). Tinkering is based on active experimental intervention, where scientists create something new to learn from, and use that knowledge to move from simplified lab experiments that isolate particular mechanisms to increasingly complex settings. Scientists intervene to isolate and test specific mechanisms, and compare divergent implications of competing explanations (Turro 1986: 885). These interventions build up understandings and inform judgements about what might be working (Hacking 1983), because scientists create patterns to learn from. The infrastructure enables the iterative generation of better explanations through repeated intervention in the world and modification of categories used to understand it.

This discovery style of research is necessary for complex innovation systems like new drugs, because the links between the aseptic conditions of the laboratory and the 'living labyrinth of human biology' (Barry 2005) are weak. Specifically, theory is a weak guide to practice, and background scientific understanding is too weak to point sharply to a solution. Especially at the frontiers of science, theories are not formulated well enough to provide explicit hypotheses. The style of research that is based on validating or confirming predictions cannot work in complex innovation, because it offers little guidance about what to do when tests fail—and tests typically fail in drug discovery and other ambiguous situations (Hacking 1983). However, discovery research does not produce simple or clear answers, so scientists need to be reflective practitioners (Schon 1983), and continually make judgements about the likelihood of different explanations for why experiments do not behave in the predicted way.

The development of the poliomyelitis vaccine illustrates the importance of an infrastructure for complex innovation systems, like pharmaceuticals. I summarize the rich description and analysis by Yaqub and Nightingale (2012). Poliomyelitis was identified as a disease in the middle of the nineteenth century, but to develop a vaccine, scientists need to understand what an appropriate immune response to the disease would be, how to stimulate it, and how to do so safely. By 1910, scientists demonstrated that monkeys that survived polio resisted reinfection, which was a proof of concept that a vaccine was possible. A scientist announced in 1911 that a remedy would be found within six months. But there was little infrastructure for the research. First, because the polio virus worked only in primates, scientists could not generate experimental data with simple animal models such as mice or even worms. Scientists had little access to monkeys who lived in India and Thailand, and in any case monkeys were very difficult to feed, care for, and deal with. Second, not enough labs could tinker with or manipulate the virus samples to see how to develop the virus into something that would stimulate an appropriate immune response, because supplies of the virus were limited. Third, an unknown variety of types of polio virus existed, so people could not compare experimental results across laboratories.

However, by that time vaccines had already been developed for smallpox, rabies, diphtheria and other diseases and saved millions of lives, without formal identification of the specific infectious agents. Polio vaccine researchers extrapolated from rabies and small pox, but the authors say that this simple extrapolation of an older operational principle produced disastrous results. In 1935, two rival teams hurried their vaccines into trial to conduct 'testing as validation' field trials, despite the lack of knowledge about the disease and what would be an appropriate and safe immune response. The trials killed or paralysed many of the 12,000 children who were vaccinated. These childhood deaths, say Yaqub and Nightingale (2012), stifled vaccine development for many years.

But then in 1947, the newly appointed director of research for a foundation started by Franklin D. Roosevelt (a polio victim since 1921) and renamed the National Foundation for Infantile Paralysis invited leading poliomyelitis researchers to conferences, and instituted round table discussions to 'encourage communication and intellectual cross-fertilization in a field notable for its lack of both' (Carter 1965: 57, quoted in Yaqub and Nightingale 2012). The National Foundation also raised significant funds and channelled those funds to specifically tackle the three impediments to vaccine development (lack of models, limited virus samples available, unknown varieties).

Polio research shifted qualitatively to a discovery style of research based on creating a better understanding of the disease and its pathology. The foundation and government agency leaders helped participants to shift from validating to experimenting in order to generate understanding that would inform scientists' ongoing judgements. They created a new infrastructure with four subsystems. First, the foundation leaders developed a collaborative commons for people to share and integrate ideas, along with a new governing structure. Second, the leaders and participants created a collective strategic focus on tackling specific problems. A large experiment spanning four universities over two years using 30,000 monkeys focused on the 'dull and menial' programme of virus typing. This experiment showed conclusively that there are three distinct types of poliomyelitis virus. Third, participants in the infrastructure integrated sciences to create the ability to innovate. They developed economies of scale by airlifting thousands of monkeys to a central facility and developing means for feeding and caring for them. They also funded tissue culturing research, including training personnel, which led to safer and faster testing cycles, more feedback among experimenters, and the ability to identify different virus strains and better experiment methods. The advances in tissue culturing allowed research groups to compare their results, which revealed new paradoxes arising from different viruses. Finally, scientists worked on particular vaccine products that would stimulate an appropriate immune response. This infrastructure all together led to a killed working vaccine in 1955, and a live one shortly after.

The authors conclude that major changes in the governance of research were necessary to accumulate knowledge, coordinate knowledge production, and govern the much more complex jumps between intermediate stages when a simple experimental model such as mice was not available. The new governance ensured that knowledge generated under local conditions could be interpreted in a cumulative manner so that a realistic understanding of whether a proposed vaccine actually works could emerge. The solution was not to throw more money at the problem or build up social concern about the disease. Rather the solution was to build the infrastructure of governing structures, strategies, tools, techniques, and experimental models, and the social practices that enabled more effective tinkering, comparisons across experimental outcomes, and knowledge accumulation.

Organizing the Infrastructure for Taking Advantage of Emergence

The polio story emphasizes the combined effects of physical technologies such as tools, techniques, and models with social technologies such as new governance structures and protocols that comprised the infrastructure. Since this book focuses on the social technologies that enable the discovery style of research for complex innovation, I summarize three sets of social technologies that I think make the infrastructure of complex product innovation capable of continually converging the fragmented and localized knowledge into viable products that actually work in the real world. These three sets of social technologies together organize and define an infrastructure for taking advantage of emergence.

Unpacking the Infrastructure into Four Subsystems:
A New Division of Labour

The first set of social technologies that enable taking advantage of emergence is an alternate division of labour based on complete processes of problem setting and solving. Complex innovation systems cannot be chopped up into discrete steps or functions. However, we can identify distinct sets of problems that need to be addressed, even though ultimately the problems are entangled. We can pull out these problems and explore them while being mindful of the whole system that they participate in and together constitute.

A particular division of labour fostered the rise of large organizations and mass production in the late nineteenth century (Chandler 1977). This nineteenth-century division of labour centred on the hierarchical decomposition of an overall process such as steel manufacturing or the mass production of automobiles into distinct steps or functions. This step-wise or functional division of labour presumes that managers understand the whole process well enough to break it up into separate steps and manage each one apart from the others. Each step or function becomes a solution, so people manage solutions, not problems. The social technology of hierarchy integrates these functional solutions into a linear flow of steps and hand-offs that afford scaled-up production, with layers of managers to oversee smooth operations and address problems arising in different subsets of steps (Perrow 1986).

Work in the infrastructure for complex innovation systems cannot be broken down into separate steps. The division of labour for complex innovation systems needs to preserve the entire process, because managers and innovators must continually evaluate and adjust the entire process, not just the steps. The new division of labour focuses on problems, not solutions, since there are no clear solutions in complex systems. Consistent with my own work on distinct communities of practice within large firms (Dougherty 2001, 2006) along with the insights from Van de Ven (1986), Van de Ven et al. (1999), and others, I focus on four basic subsystems of discovery: innovation project management (e.g. the live virus solution for polio), innovation process or knowledge management (e.g. identification of distinct strains of the virus), strategic management (e.g. focusing resources on overcoming specific constraints in developing polio vaccine), and institutional management (e.g. developing governing structures that guide the world-wide collaboration for the polio vaccine).

Each of these subsystems brackets out different aspects of the complex product innovation to be discovered and developed, and so helps to emphasize the importance of working on each problem. This division of labour fosters taking advantage of emergence because each of the four problems must be addressed at the same time. This division of labour also disentangles the Gordian Knot of complexity into more tractable sets of innovation activities that people can carry out somewhat separately, provided they continually adapt to feedback about work in the other subsystems. The interdependence of these subsystems also leads to integration not by hierarchy, or one to many links, but by heterarchy, or many to many links (Ansell 2011). Heterarchy involves many relations that are comprised of shifting configurations of the same people who over time form different groups that come together for particular challenges. Heterarchy enables both positive and negative feedback, and pulls the overall system towards a bounded instability (Plowman et al. 2007).

That each subsystem of problems must be addressed in its own way became apparent during our research. We noticed differences in perspective among scientists working on drug projects (project scientists), scientists working with enabling technologies such as genomics or bioinformatics (knowledge system scientists), and strategic managers

(Dougherty and Dunne 2012; Dougherty et al. 2013). The differences were like the interdepartmental thought worlds that I developed years ago to explain why the departments involved in new product development did not always work well together (Dougherty 1992). Each thought world knows different things about developing new products, and also knows things differently. However, the thought worlds in pharmaceuticals are each vast networks of people working around the world in different institutions and organizations. They are not located inside individual firms, which is why the new division of labour is needed.

Our academic papers explore specific tensions between two of these different groups that seemed to suppress emergence. For example, knowledge system scientists seem to think that project scientists are not working rigorously, so some try to replace local tinkering with massive 'rational' assays. Strategic managers seem to think that project scientists explore too many interesting questions rather than get to the right answers, so they push for sharper decisions based on clear information. Project scientists are frustrated by the lack of support and the long time it takes managers to allocate resources, so they push forward on their own, often myopically.

I finally figured out that each group in pharmaceuticals was working on a different problem of discovery in the complex innovation system, and each group needed to surface and develop a different configuration of relationships among different elements. The tensions we saw resulted from limited appreciation for the existence of all four problems in complex innovation, so everyone focused on innovation projects. People did not pay enough attention to the problems of integrating the explosion of new sciences, creating alternate business models, and connecting various institutions in the overall infrastructure to enable more effective collaborations. It would be as if the polio infrastructure did not try to develop tissue cultures that all labs could use to experiment with vaccines, and did not try to channel research strategically so knowledge of a viable vaccine could accumulate.

I argue that all four subsystems need to fully address their distinct problems of discovery in their own right, and that all four need to continually integrate by mutually framing and constraining each other. Addressing all four different problems reflects what I and others had discovered for innovation on a much smaller scale within firms. Van de Ven (1986) emphasized four central problems in innovation: project, technology, business, and institutional. My early research found that product innovation teams worked apart from the rest of the organization (Dougherty 1992; Dougherty and Heller 1994; Dougherty and Hardy 1996). The innovation teams received little help with developing the technological infrastructure that would enable them to create their new product. For example, a new product team in a heavy equipment manufacturer had to develop general hydraulic technology before they could create their particular product, even though the firm's entire product line would benefit from hydraulic technologies. A team at a food company had to work with outsiders to develop the manufacturing technique their new product required, even though this technique became the industry standard a few years later. These companies were not developing long-term technological capabilities or innovation strategies to support ongoing product innovation. Instead, everyone except innovators worked on keeping the existing functions going as is. Product innovation was like passing a corporate kidney stone.

While some large organizations still tend towards this hierarchical or bureaucratic organizing, innovative organizations have learned to develop all four subsystems of innovative problem setting and solving (Adams 2004; Dougherty 2006). The pharmaceutical industry faces the added challenge that their subsystems of innovation are all vast, interorganizational networks. Being an innovative organization in a complex innovation system requires much more system-wide development. I outline the four subsystems of complex innovation here.

THE INNOVATION PROJECT SUBSYSTEM
The innovation project subsystem of complex innovation addresses the discovery problem of actually building the product, by ferreting out the many specific components, figuring out how these may go together to generate some desired functionality, and exploring innumerable possibilities as effectively as possible. Pharmaceutical project teams seek to develop a viable configuration of interdependencies among the disease processes, a molecular compound, and the rest of human biology that would comprise a good drug product. Denrell et al. (2004) suggest that dealing with complex problems is like navigating in a labyrinth. The central challenge for this activity: are we navigating as well as we can in this multi-dimensional labyrinth of complex new products?

For drug development as well as for other kinds of complex innovation, the work of project innovators is the heart of the innovation. In drug discovery, multi-functional teams comprised of biologists, chemists, physiologists, and other experts seek to synthesize a molecule that can enter the body, travel through the blood stream to the disease site, and ameliorate a particular disease (Dunne and Dougherty 2016). The work is very hands-on, very concrete, and very embodied in the full sense of scientists working with the physical materials in their labs, as described by Knorr Cetina (1999). Project scientists are uniquely able to see how particular patterns of elements might work and explicate how mechanisms function. In other complex innovation systems, multifunctional project teams work in a similar manner to discover and create the product or service that will, for example, address a lack of job skills, create a particular approach to renewable energy, carry out community policing, and so on.

THE INNOVATION KNOWLEDGE MANAGEMENT SUBSYSTEM

The innovation knowledge management subsystem of complex innovation addresses the discovery problem of integrating diverse knowledge to support projects. Like those in the polio infrastructure who developed tissue culturing and animal models, knowledge scientists in pharmaceuticals seek to discover viable configurations among sciences and technologies that can provide better testing regimes, better models, and enhanced project searching. The central challenge of this subsystem: are we integrating all the knowledge well enough to enhance new product or service development? That is, are we helping to navigate in the labyrinth of complex innovation projects by shaping choices and identifying good alternatives.

I label this subsystem designing the strategic path for project innovation, because new sciences support general questions that span therapy areas, such as how well do compounds bind or are they efficacious? Pisano (2006) argues that integrating the new and old sciences is the only way to deal with the profound and persistent uncertainty in this innovation system, because there is no way to sort out what may be happening by looking at just one aspect of the problem, or at each aspect in isolation. Others also emphasize collaboration among scientists and agencies (Powell et al. 1996). Despite the emphasis, new sciences and technologies have not been integrated adequately into the drug discovery process, so this innovation system is not using all the knowledge available to it (Dougherty and Dunne 2012). Other complex innovation systems such as education or economic development have even more knowledge available from the behavioural, economic, and social sciences, but it seems to me they integrate this knowledge even less effectively than pharmaceuticals. Participants in the pharmaceuticals infrastructure at least acknowledge the need to integrate all these sciences, but other complex innovation infrastructures in my view seem unaware of the need to integrate knowledge.

THE STRATEGIC MANAGEMENT SUBSYSTEM

The strategic management subsystem in the complex innovation infra-structure addresses the discovery problem of continually creating value by finding viable configurations of interdependencies among knowledge resources and value-creating opportunities. Knowledge is the primary resource in complex systems, so the strategic management subsystem continually sets and solves the problem of whether we are leveraging knowledge resources available to us as effectively as we can. This subsystem searches for better ways to leverage the knowledge that is available. 'Strategic management' does not refer only to business managers inside firms, but rather to the strategic practice of defining objectives and then marshalling resources and orchestrating activities to get to there, from here—including rethinking the paths as the infra-structure proceeds. The polio story highlighted the essential role of strategic management on the part of the foundation director, advisory boards, research lab managers, and government and science leaders around the world, who worked strategically to make thousands of animal models and refined tissues that carried the disease available to scientists around the world.

However, ambiguity leads to very long product development cycle times, so the strategic subsystem cannot concentrate on short-term adaptations to changes. Strategic management must muster the staying power to persist and learn by enabling business models and value-creating opportunities for deploying innovations to continually emerge (Lynn et al. 1996). Strategic management develops future possibilities that emerge over time. To do so, they must develop and foster different kinds of temporal structures that can map further out in time.

THE INSTITUTIONAL SUBSYSTEM OF COMPLEX INNOVATION

The institutional subsystem of complex innovation addresses the dis-covery problem of orchestrating the many organizations, institutions, regulators, and other agents that each generate essential knowledge for innovation. This subsystem continually creates configurations of inter-dependencies among relational elements such as goals and processes that would enable ongoing collaborative commons. The collaborative commons enable strategic managers to create new value-creating oppor-tunities; they enable knowledge scientists to integrate their diverse capabilities into new tools and techniques; and they enable project scientists to create viable new drugs. The central challenge in the ecol-ogy subsystem is: are we using society's resources (all that knowledge and all that investment in infrastructure) well? To use society's knowledge resources well, the institutional subsystem has to promote the long-term co-evolution of, in the case of pharmaceuticals, science, technology, clinical care, industry, and regulations.

Institutional arrangements that support innovation are increasingly vital for many industries, because competitive pressures and rapid technological changes reduce the ability to rely on in-house R&D (Chesbrough 2003). Pharmaceutical innovation has no choice but to collaborate across many organizations and agencies, because many public and private R&D organizations around the globe create basic science, including universities and government labs, alliances, and other consortia. As well, government offices such as the FDA in the US regulate public safety and welfare so they too need a seat at the collaborative table.

The division of labour for complex innovation systems into four subsystems is both familiar and not so familiar. On the familiar side, these subsystems already exist for pharmaceuticals and for many other complex innovation systems, so we can ask if participants in each are addressing their distinct problems well. The answer for pharmaceuticals is no. People do not adequately integrate their scientific knowledge into strategic paths for drug discovery. Many firms are still stuck in a blockbuster business model and the infrastructure overall does not generate enough value-creating opportunities to use all the innovations that can emerge. Institutional arrangements are rigid and risk adverse.

The division of labour is also unfamiliar, because it does not include levels or hierarchy. Each of these subsystems is a very large network of numerous actors and institutions. Developing a new drug therapy for cancer or Alzheimer's, for example, involves thousands of people and agencies, so higher levels do not subsume innovation projects like this. In addition, each subsystem is an active, ongoing set of identifiable and observable processes and practices that are carried out by identifiable and observable people. None of the subsystems is an abstracted force. All are active, concrete activities. Moreover, the subsystems are entangled, with each one connected in diverse ways to each other one. Complex systems cannot be decomposed or even considered as 'nearly decomposed' (Simon 1977).

*Abductive Learning Routines that Animate and Motivate
the Infrastructure*

The second set of social technologies for infrastructures of complex innovation systems build abductive reasoning into learning routines.

Abductive learning routines animate the processes of problem solving in all four subsystems, and provide a common process that facilitates ongoing interactions among the subsystems. Product innovation by definition is the creation, combination, and recombination of knowledge into novel configurations, so all innovation concerns collective learning and collective configuring. Abductive learning routines facilitate taking advantage of emergence and so provide the approach to knowledge creation, combination, and recombination that works in infrastructures of complex systems of innovation.

The central research paper from our study of pharmaceuticals details the abductive learning routines used by project scientists (Dunne and Dougherty 2016). Prior to our study of innovation in pharmaceuticals, I had interviewed more than 300 people in large firms who were working on new products. But the approach to learning described by the drug discovery scientists confused and surprised me and my colleagues. The scientists said they followed clues, focused on figuring out the right questions to ask, engaged with each other to make sense of experimental results, and relied on luck and intuition. Other innovators that I had interviewed before this study were working on more incremental innovations, and so could drive to solutions that they knew existed. We eventually realized that the discovery scientists were describing how they deal with emergence. We came to understand that there are regularities in how the drug discovery scientists learned, and that they were talking about routines for emergent learning. The rest of this book explicates the abductive learning routines across all four subsystems, and describes how they are enacted in different subsystems of the infrastructure.

LEARNING ROUTINES

I begin this discussion of the social technology of abductive learning routines by explaining the term 'routines'. Academics define this term in different ways, so routines are one of the contested concepts in organization studies. I think that defining learning routines based on Feldman and Pentland's theory leads to a better set of social technologies for taking advantage of emergence. According to Feldman and Pentland (2003), routines are recognizable, repetitive patterns of interdependent actions. But unlike other definitions, these routines are not rigid standard operating procedures or recipes that are carried by rote, in a de-contextualized or abstracted manner. Instead, routines emerge over time, because they involve multiple actors with divergent goals and understandings, and are carried out in a variety of unique situations. People adapt their routines to the particular contexts they are in and to the actions of the people with whom they are working, so routines continually absorb these contextual variations. Feldman and Pentland's (2003) definition captures both the structuring that routines perform and the situated, everyday practices through which people continually create and enact their shared social structures. Routines enable emergence.

Learning routines provide essential social technologies for the infrastructure of complex innovation in two ways. First, learning routines structure the everyday work around emergence. Learning is systematic, methodical, and deliberate in complex systems, not random, ad hoc, or idiosyncratic. Structuring orders all the possible actions into a more coherent understanding of this is what we do here, how we do it, and how we work together—structuring makes collective work sensible and doable. Second, the Feldman/Pentland idea of routines emphasizes what people actually do, and so captures the 'performative' nature of everyday work in complex systems. As Tsoukas and Dooley (2011: 731) explain, performative means that people are not presented with objective problems. Rather, people help bring those problems forth, or enable them to emerge, 'through the application of the symbols, categories, labels and assumptions contained in the tools they use and the practices they draw upon'.

Cohen (2007) associates the performative idea of routines with Dewey's pragmatism-based perspective on habit. According to Dewey, habit is a learned predisposition to ways or modes of response, not just particular acts. Habit, according to Cohen (2007), is like a skill that people learn over time, like the skills of a craftsperson or a master violinist. I apply this idea to learning routines and suggest that abductive learning routines provide a disposition to action, not an automatic or robotic response. Abductive learning routines enable people to do things collectively but also adjust readily to the messiness of emergence. Pentland et al. (2011) demonstrate empirically that even a presumably standard routine such as one for processing invoices changed continuously. People do not socially construct a precise replication of the activity. Rather, they socially construct the meanings and actions that are necessary to create and stabilize the situation.

A brief example helps to explicate the socially constructed nature of routines. Some make fun of the idea of social construction by asking if we want to fly across the Atlantic in a socially constructed airplane, as if the term means that people just make things up. Social construction actually refers to the artful competence of the reflective practitioner who can deal with surprises by fitting routines to the situation. In the case of flying across the Atlantic, we hope the airplane has been well constructed physically. But we also hope that the pilots are able to respond immediately to sudden surprises such as wind shear, thunderstorms, or an engine shutdown. We want the pilots to socially construct, on the spot, a solution by drawing on their training and experience to use their routines for recovery in a way that fits this situation. We want them to be disposed to take effective action, not to respond in a mindless, rigid manner.

However, people and the social context can reinforce rigid and mindless rather than flexible and thoughtful application of routines. If this dark side of routines unfolds in the infrastructure of complex innovation systems, innovators will not be able to take advantage of emergence.

Many studies of knowledge workers find that they readily deal with ambiguous task situations, and work with incomplete knowledge (Barley 1996). People do not require simplification to work productively. Engineers, copy repair technicians (Orr 1996), information technology (IT) technicians, and navy personnel on an aircraft carrier (Weick and Roberts 1993) work through complex problems despite ambiguity, using their routines flexibly to adapt to the situation. The work of scientists also embraces emergence, because according to Grinnell (2009) 'the path to discovery in everyday practice is ambiguous and convoluted, with lots of dead ends. Success requires converting those dead ends into new, exciting starts.' Other studies of scientists also highlight their highly situated, tacit, and embodied practices of knowing (Knorr Cetina 1999; Latour 1987).

ABDUCTIVE LEARNING ROUTINES

Abductive learning routines have a particular character that supports taking advantage of emergence, because abduction is the logic of discovery. Abductive learning routines directly fit the empirical phenomenon of drug discovery and other complex innovations, because they explicitly address problems where the knowledge is limited, incomplete, and fragmented. Our central paper on innovation in pharmaceuticals develops the idea of abductive reasoning to explain the approach to learning that the scientists in our study follow (Dunne and Dougherty 2016). Our research generates a new and richer understanding of this form of reasoning, and demonstrates how abductive reasoning addresses the specific challenges of creating new products in complex innovation systems. In this book, I extend these findings to all four subsystems of discovery in this complex innovation system. I show how innovators use abductive learning routines to take advantage of emergence. Abductive learning routines promote the discovery style of research that allows scientists to tinker, to accumulate diverse insights into new understandings, and to fine-tune better alternatives.

But first, definitions are in order because the few management academics who use abduction define it differently. As well, the majority of management academics who do not know about abduction may find the term to be alien. I define abduction as the logic of discovery, based on Peirce, the nineteenth-century logician who developed the idea. Abduction comprises the deliberate reasoning that leads to scientific discoveries and learning the secrets of nature (Nesher 2001). Repeating Locke et al.'s (2008: 907) quote of Peirce from the introduction to the chapter: '[d]eduction proves that something *must* be; induction shows that something *actually is* operative; abduction merely suggests that something *may be*' (emphasis in original). Weick (2005) describes the abductive process as 'clues giving rise to speculations, conjectures, and assessments of plausibility rather than a search among known rules to see which ones might best fit the facts' (quoted in Locke et al. 2008: 906). Simon (1977) also discusses the abductive process of discovering laws in raw data based on pattern recognition and abduction of hypotheses on laws that may regulate observed patterns. We understand abduction to mean the creation of novel hypotheses about what might be going on in phenomena that are not fully understood. Abduction 'is the process of reasoning in which explanatory hypotheses are formed and evaluated' (Magnani 2001: 18).

According to Peirce (cited in Nesher 2001), only the abductive processes of formation (discovery) can supply new concepts and rules which later are evaluated and selected inductively. Peirce says that deduction evolves the necessary consequences of a pure hypothesis, but depends on abduction to suggest a hypothesis in the first place. Induction evaluates the deduction with observational data. Again according to Peirce: 'induction is the experimental testing of a theory. . . . The only thing that induction accomplishes is to determine the value of a quantity. It sets out with a theory and measures the degree of concordance of that theory with fact. It can never originate an idea whatever' (Nesher 2001).

We use the term 'abductive reasoning' to encompass formulating, evaluating, and reframing hypotheses about possible new products. I extend these ideas to include formulating, evaluating, and reframing possible new science and technology systems that enable drug discovery, possible value-creating opportunities that deploy and apply new drug products, and possible new institutional arrangements that enable new modes of interactions. Innovators in all subsystems cycle through these three abductive learning routines of formulating, evaluating, and reframing their hypotheses or models continuously. I will bring this cycling alive with particular social processes that I will explicate in Chapter 2.

However, to the majority of academics the idea of abduction is alien in three ways, which get in the way of recognizing, let alone adopting, this form of reasoning. First, the dictionary meaning of abduction is kidnapping, or wrongful carrying off of a human being. In popular use, aliens carry off or abduct people from earth to their space ship, according to certain kinds of newspapers. Peirce's use of abduction reflects the sense of taking away as well, but he means away from existing expectations to something new and unexpected, not away from earth to a space ship.

Another definition that makes abduction alien in organization studies is that some conflate it with induction. Ketokivi and Mantere (2010) refer to abduction as induction. Arthur (2014) explains that under Knightian or fundamental uncertainty events take place in the future so outcomes are unknowable and decision problems are not well defined. To deal with such situations, he explains that people engage in surmising, using past knowledge and experience, and using their imaginations to try and come up with some picture of the future, and then proceed on that basis. While Arthur's description reflects abduction, he calls this process induction. More generally, when I mention the idea of abduction, some say 'isn't that induction?' Peirce differentiates abduction as the logic of discovery from induction as the logic of evaluation. If these two forms of logic are not distinguished, then induction goes from the particular to general without evaluation, and we have no logic of discovery.

Third, and perhaps most problematic, ideology about proper research renders abduction alien, too. The logics of abduction and induction are based on plausibility and probability respectively, not on absolute truth, and so are inherently incomplete. Incompleteness makes it very difficult to assess scientific claims, as Ketokivi and Mantere (2010) explain. The management academic domain emphasizes deductive logic, I think, for this reason—reviewers have a much easier time assessing deductive claims with the normative rule of logical coherence (Ketokivi and Mantere 2010: 316). Provided the theory that one deduces hypotheses from is logically complete, reviewers and peers can assess claims based on how clearly the researcher goes from axiom to conclusion. Deduction presumes the strict condition that the conclusion must follow analytically from the premises. According to Peirce, deduction is the only necessary reasoning. It starts from a hypothesis, the truth or falsity of which has nothing to do with reasoning; and of course its conclusion is merely ideal (cited in Nesher 2001).

However, if the premises from which specific conclusions are drawn are themselves incomplete, deduction cannot be logical. According to Arthur (2014), under Knightian uncertainty, decision problems are not well defined so pure deductive rationality is not well defined either. There cannot be a logical solution to a problem that is not logically defined. Grounded theory-building and other forms of qualitative research in the social sciences rely on abductive logic to discover and elaborate on theories about managing and organizing, because existing theories are incomplete, at best. Unfortunately, deductive methods dominate our journals and doctoral programs, and qualitative studies are much less likely to be published. I find it ironic that deductive researchers sneer at abductive research for not being rigorous, and instead emphasize their illogical conclusions. We need to learn to rely on both kinds of science to enhance our claims.

Organizing for Heedful Interrelating

One last set of social technologies underpins the infrastructure of complex innovation systems. This set organizes by defining and shaping roles and relationships among many people, agents, large and small firms, public and not for profit organizations and associations, and regulators so they can collectively take advantage of emergence. I suggest that a social context of heedful interrelating (Weick 2005) fosters abductive learning routines. Heedful interrelating and abductive learning are both rooted in the pragmatic understanding of human intelligence as creative activity, or the improvisational response of human beings to the concrete situations in which they are implicated (Joas 1996). In pragmatism, thinking and doing are two sides of the same coin of intelligent being in the world. Herbert Dreyfus explained that, according to Heidegger, human intelligence is not based on the ability to store massive amounts of information. Rather, our intelligence arises from our ability for *being in the world*, and for figuring out what information is relevant to the particular situations we are in. The classical pragmatists portray human beings as situated actors, and as such, as creatively responsive beings (Colapietro 2009). Creative activity, or the improvisational responses of human beings to the concrete situations in which they are implicated, is the most basic form of human action, according to Joas (1996).

Organizing reflects the doing side of thinking and doing, and so is an inherent part of abductive learning routines. However, I pull the idea of organizing out of the entangled process to clarify and elaborate on its special properties. My reason for doing so is that people seem to ignore the ongoing collective processes of interacting, and focus instead on individual processes like creativity or cognition, and on heroic individuals who invent new ideas. Organizing to enable the collective thinking and doing gets lost, or pushed into the background.

I base the social technologies for organizing infrastructures of complex innovation on pragmatic ideas that are reflected in Karl Weick's idea about collective mind. The social context matters a great deal for collective learning according to Weick (1995, 2005), who builds on Hutchins's (1995: 176) idea that 'the cognitive properties of human groups may depend on the social organization of individual cognitive capabilities'. A pragmatics-based view of the social organization of cognitive capabilities has seeped into organization theory as well, and lays the groundwork for the new social technologies for organizing that I develop. For example, Corley and Gioia (2011) draw on pragmatism to say that making our theories more relevant to practice would make them better. In pragmatism, meaning (theory) is linked closely to action and depends on hands-on experimenting that is based on the confrontation with real problems (Ansell 2011). Corley and Gioia summarize a number of organization theorists who view knowledge as a recursive dialogue between practice (action) and meanings (theory). Tsoukas and Knudsen (2005) also explain that social practices and meanings are mutually constituted. A person's understanding does not reside in his or her head, but in the practices in which he or she participates. They say that social activity, not the cognizing subject, is the foundation of intelligibility.

Cohen (2007) suggests that pragmatic thinking is reflected in the practice-based view (i.e. knowledge and its value emerge from ongoing and situated actions of organization members (Orlikowski 2002)). Cohen (2007) also says that Weick's ideas about mindfulness and collective mind reflect pragmatism. Weick (2005) argues that mindful organizing overcomes the limiting effects of what I call the old social technologies for organizing (hierarchy and decomposition into separate steps). Mindfulness involves the ongoing creation of new categories out of the continuous streams of events that flow through activities, and a more nuanced appreciation of the context of events and of alternative ways to deal with that context. The collective mind enhances the intelligence of the social system (Weick and Roberts 1993).

According to Weick (2005), collective mind 'can be conceptualized as a kind of capacity in an ongoing activity stream that is emergent and takes different forms depending on the ways in which the activities are interrelated'. He quotes Taylor and Van Every (2000: 207):

> Groups composed of individuals with distributed ... partial ... images of a complex environment can, through interaction, synthetically construct a

representation of it that works; one which, in its interactive complexity, outstrips the capacity of any single individual in the network to represent and discriminate events...Out of the interconnections, there emerges a representation of the world that none of those involved individually possessed or could possess.

Weick and Roberts (1993) conceptualize collective mind as patterns of heedful interactions in a social system. Heedfulness refers to the way behaviours are assembled—carefully, creatively, purposefully, and vigilantly. With more heed, people interrelate their actions with more care. Heedful interrelating means that people construct their own actions (contributing) while envisioning a system of joint action (representing) and interrelate their action with that of others (subordinating). The actions of one person thus begin to converge with, supplement, assist, and become defined in relation to the imagined requirements of joint action, but only when the joint situation is also represented in the actions of others. People construct and reconstruct their interrelations continually through their ongoing activities of contributing, representing, and subordinating (Dougherty and Takas 2004).

According to Weick and Roberts (1993), the more heed that is reflected in a pattern of interactions, the more developed the collective mind, and the greater the capacity to comprehend unexpected events that evolve rapidly in unexpected ways. With heedful interrelating, people enact aggregate mental processes that are more fully developed than those found in efficiency organizations. Complex patterns can be encoded by patterns of activation and inhibition among simple units, if these units are richly connected. Overlapping knowledge allows for redundant representation that enables people to take responsibility for all parts of the process to which they can make a contribution.

One might ask: how can we generate such a seemingly complex pattern of interactions so that people in diverse agencies, organizations, and institutions can share ideas, co-create new understandings, and accumulate knowledge systematically by attending to feedback about the actions of others in the infrastructure? One answer is that many of us already work this way. Studies show that knowledge workers of all stripes actively and readily interact heedfully as they collectively generate, accumulate, and apply emergent knowledge. A famous ethnography of copy repair technicians details how these 'lowly' workers figure out how to overcome complex machine failures that the Ph.D.s said could never happen in the first place. The technicians interacted continually with each other, sharing war stories that embodied extensive experiences with similar challenges, and attended closely to the context in which the machine operated for clues to possible sources of breakdown (Orr 1996). Other research shows that engineers readily help each other by trading knowhow without giving away company secrets (von Hippel 1988).

Scientists also interrelate heedfully as they seek to discover new understandings. According to (Grinnell 2009: 14): 'In everyday practice, discovery begins in a community. Community offers continuity with the past and interconnectedness of the present. Each researcher or group of researchers initiates work in the context of prevailing experiences and beliefs.' He explains that people overcome subjectivity through inter-subjectivity, or reciprocity of perspectives. Knorr Cetina (1997) argues that science and other professional knowledge work generates a new kind of sociality. Building on Rheinberger (1992), she notes that 'epistemic things' are the scientific objects of investigation at the centre of a research process. Unlike static instruments and tools, these epistemic objects are continually unready to hand, or continually unfolding structures. Because of the continually unfolding nature of these epistemic objects, knowledge workers are drawn strongly to them, and this deep attachment creates social bonds. The collective pursuit of knowledge binds people with a common identity that is centred on the knowledge object, and a mutual recognition and sense of belonging.

While heedful interrelating typifies knowledge work, I recommend a few social rules to bolster our predisposition to work heedfully. Rules and resources constitute structure and organizing. One rule is that leaders need to foster heedful interrelating, and to encourage people to interact in this way. Weick and Roberts (1993) suggest that leaders can foster heedful interrelations by making heedfulness visible, rewarded, modelled, and preserved in vivid stories so newcomers can learn it. Through mindfulness, people attend to failure, avoid simplicity rather than cultivate it, are as sensitive to operations as much as to strategy, and organize for resilience. The second rule is to emphasize innovation, which refers to the continued development of new products and services that resolve critical problems in the infrastructure. Innovation does not mean creating novelty for its own sake. The third rule is to both recognize and deliberately foster all four subsystems of problem setting and solving in the infrastructure for complex innovation.

I develop particular ways to promote heedful interrelating for each of the subsystems in the next four chapters. However, my basic argument is that people do not find it difficult to organize for heedful interrelating. Knowledge professionals tend to work this way, and heedfulness fits with our ability for being in the world and reacting to actual situations intelligently. Unfortunately, mindless organizing often dominates. Mindless organizing, Weick (2005) tells us, attends to success, simplicity, and hierarchy. Heedlessness is a failure of intelligence, a failure to see, to take note of, or to be attentive to. Heedlessness is not a failure of knowledge. When boundaries of an envisaged system are drawn more narrowly, subordination becomes meaningless. Attention is focused on the local situation rather than on the joint situation. Organizing with a hierarchy divides labour into separate steps or functions and so bounds the envisioned system very narrowly, since people only see their particular step. If the system encourages individualism, survival of the fittest, and macho heroics, people will neglect the heedful practices of jointly representing the shared situation and subordinating their actions to those of others. If people do not act heedfully with respect to others, interrelating becomes careless, and key people and activities are overlooked. When organizing loses collective mind, events become incomprehensible, contributions become more thoughtless and less interdependent, people become isolated, the system pulls apart, and problems become more incomprehensible.

The social technologies for organizing complex innovation require a different but not unfamiliar perspective. The social technologies for organizing that emerged to support modern industrial society—namely bureaucratic organizing and command and control leadership—are still fostered, however inadvertently, in part because we do not know how else to organize. But these old social technologies for organizing suppress abductive reasoning and heedful interrelating. Second, our society fixates on the individual, not on the social context. At least in the United States, any time complex innovation systems stumble (e.g. the Veteran's Administration), we install a new leader and then walk away, as if this one person can use his X-ray vision to set the complex system on the right trajectory. Leadership is vital for complex innovation systems, but the leaders' job is to develop and maintain heedful interrelating so that people collectively take advantage of emergence. Third, we tend to think that people are pretty dumb and need to be directed at all times, and that they are guileful, self-centred, and lazy rather than responsively creative. As I have already noted, research shows that knowledge workers, from top scientists to lowly technicians, do indeed work creatively and are adept with the emergence of knowledge (Barley 1996). We are certainly annoying, and we definitely need some social structuring to direct our energy and attention to collective learning. But we are intelligent enough to respond to situations surprisingly well, if given the chance.

Conclusion

We come full circle to the infrastructure for taking advantage of emergence. Abductive learning routines cannot lead directly to clear scientific claims, because every logical sequence of abductive reasoning has only conditional and relative factual capacity. However, infrastructures for complex innovation exist precisely because it is difficult to assess scientific claims when the knowledge is incomplete and fragmented. According to Pavitt (1999), the outputs of science are problem-solving skills and infrastructure, not information, because nature is inherently unpredictable. Society invests substantial public and private resources in an infrastructure to exploit research, to innovate, to import technology, to access international science, and to make sense of incomplete knowledge.

Abductive learning routines along with heedful interrelating facilitate taking advantage of emergence in infrastructures of complex innovation systems in four ways. First, abductive learning underlies the discovery style of research that Nightingale (2004) argues to be necessary for a good infrastructure. Abductive learning routines enable innovators to leap from fragmented information to informed intermediary models for ambiguous phenomena, and these models enable innovators to impute the value of their actions going forward (Denrell et al. 2004).

Second, abductive learning routines animate the learning within and across all four subsystems. They provide the same logic for working on different but interdependent problems of discovery. Using this same logic, people can work in a distributed manner pursuing their own discovery problems based on local knowledge, but adapting continuously to feedback about the actions of others. 'Order' in the form of viable innovations emerges from the self-organizing processes of participants in the infrastructure, provided they work heedfully.

Third, by actively working on all four sets of innovation challenges with abductive learning routines and by continually fostering heedful interrelating around the collective objectives of solving the innovation problems, the infrastructure handles safety and risk. Safety is an emergent property of the entire infrastructure (Leveson et al. 2009). If people in all subsystems are attuned to spotting perturbations and to thinking about how their patterns interweave, potential risks and safety problems are much more likely to be noted and addressed. And by cycling through the three learning routines for formulating, evaluating, and reframing hypotheses about innovation in all subsystems, people can interrogate their hidden assumptions and look for unexpected connections. But without this process of reasoning, people avoid implementing innovations in complex innovation systems because they can be risky (Nembhard et al. 2009).

Finally, cycling through abductive reasoning enables the infrastructure to persevere in accumulating knowledge. In the next chapters, I explain how innovators can integrate the fragmented knowledge available to them into a model or hypothesis of the most essential interdependencies among elements that might form a product, process, strategy, or governing structure for collaboration. Innovators then evaluate their hypothesis of interdependencies by exploring its consequences, and figuring out how unexpected consequences can be useful. Innovators use what they learn to refine or reframe their hypothesis of product, process, or strategic possibilities, and cycle again through the discovery, evaluation, and refinement process. Throughout, innovators build on their own background knowledge and rely on the wisdom of others who are also 'navigating in the labyrinth' of the complex innovation system.

2

Abductive Reasoning as the Foundation for Taking Advantage of Emergence

The purpose of this chapter is to explain how and why abductive learning routines enable scientists and other knowledge professionals to take advantage of emergence as they innovate, in all subsystems of discovery across the infrastructure of complex innovation systems. I detail the abductive learning routines for formulating, evaluating, and reframing hypotheses that lead to the discovery of a viable new product and to the discovery of viable knowledge systems, business approaches, and institutional arrangements that will support product innovation.

Danielle Dunne and I discovered the project-level abductive learning routines through grounded theory building with our interviews with drug discovery scientists. From the beginning of our research, we thought that the scientists were working systematically somehow, but in a way that deviated from the usual linear approach to problem solving. Their systematic processes of inquiry also seemed to clash with the processes for validation, decision-making, and regulation throughout the infrastructure, and the clashing inhibited the scientists' ability to use their systematic approach to innovation. Focusing on the project level in the infrastructure, we worked to understand this systematic but unfamiliar way of figuring out complex problems. Typical of grounded theory building, we iterated from our data to existing theories about learning under complexity and organizing for innovation, and back to interviews with more scientists. We also presented preliminary ideas at conferences and seminars, and several different colleagues said that we seemed to be talking about abduction. We iterated again between our data and literature about abduction, and realized that abductive reasoning is the basis for the systematic approach to complex innovation that we observed in our data.

I extend the project abductive learning routines that we developed (Dunne and Dougherty 2016) in two ways to create a new framework for taking advantage of emergence throughout the innovation infrastructure. First, I show how abductive learning routines and heedful interrelating fit well with the literature on how scientists actually work, on learning under complexity, and on organizing for innovation. The close fit indicates that scholars have been continually arriving at similar ideas. I hope that my synthesis of all these good ideas pushes us past continual rediscovery and on to applying these important ideas to addressing important problems. My second extension is to show how the three abductive learning routines also enable taking advantage of emergence for all four subsystems in the complex innovation infrastructure, and I develop these ideas in the next four chapters.

What is Complexity and Taking Advantage of Emergence?

First, I briefly review complexity and what it means to take advantage of emergence. I use 'complexity' to refer to the complexity of the new product and of the task of constructing that product. This product and task complexity highlight the lack of understanding about causes and effects and about what relevant knowledge might be for the product, as well as the complexity of the 'real world' that the product needs to work reliably in. In the case of new drugs, this complex real world includes very limited knowledge about diseases and the human body, and complexities in the health care system, the business environment, and the regulatory environment. It also highlights institutional complexity (academia versus industry, science versus business, managerial control versus technical understandings), and temporal complexity (quarterly financial reporting versus unpredictable product cycles that average thirteen years).

I find that Snowden and Boone's (2007) distinction between complicated and complex usefully sorts out what we often mix together. Applying their ideas to innovation, they would suggest that innovators working on complicated problems can discover relationships between cause and effect because these relationships exist, even though they are not immediately apparent to all. These innovators need to investigate multiple options and there may be multiple 'right' answers. Answers do exist, but they are hard to find and it is difficult to optimize single solutions. Most of the innovations in software, robotics, and electronics are complicated rather than complex, because innovators work with existing architectures that define relations among elements. Unfortunately, most studies of 'complex' innovations concern software, robotics, and electronics, and so are not about complexity at all. Innovators working on complex problems face unknown relationships between cause and effect, and no answers exist. Innovators working on complex problems must discover and indeed create knowledge about relationships between cause and effect that matter to their innovation. As well, any major change introduces unpredictability and flux.

Emergence explains how answers might spontaneously arise from the actions and interactions among people and subsystems. For my purposes, emergence means that problem definitions and problem solutions arise spontaneously from the circumstances, without any intervention from controllers or managers. In complex systems, outcomes are unknown because they take place in the future. Innovators cannot gather and process information ahead of time, because what may be relevant is unknown and must be discovered. As well, the information is abundant but noisy, fragmented, and locally embedded in a variety of places. To take advantage of emergence, innovators need to detect useful information despite the noise, and gather up information in the process of innovation. They need to interpret possible alternatives and their consequences by searching far from familiar understandings, but also keep the whole in mind since they do not know what the necessary elements are or how those elements might interact. Taking advantage of emergence means that innovators actively try to surface possible answers by continuous, concrete problem solving, and by staying ready to spot even minor perturbations that may escalate into major problems—or opportunities.

To illustrate both the challenges of taking advantage of emergence and why it matters to be able to do so, I paraphrase Pisano's (2006) example of complexity in pharmaceuticals. Pisano (2006) illustrates complexity in drug discovery with the example of a cancer drug. Consider a drug candidate for cancer in early-stage human clinical trials. It is a novel compound that targets a receptor on the surface of the cancer cells. This receptor is hypothesized to play a critical role in the cascade of biochemical reactions that lead to uncontrolled cell growth. However, despite the fact that the active ingredient significantly reduced tumour size in laboratory animals, the early results in the human trial are disappointing. A small percentage of patients seem to respond well; they get a very significant reduction in tumour size that appears to be durable. Another group of patients gets a temporary reprieve. But unfortunately many patients seem to get no benefits. What is going on?

There are many possible explanations, but the information that is required to solve this puzzle is likely to come from various sources. Maybe the receptor does not have the role hypothesized. Maybe interfering with this receptor has no impact at all on cancer growth. Maybe the effect of this receptor on cancer depends on what is happening simultaneously with several other receptors. The human life system is very redundant, so perhaps turning off this entry into the cell triggers the opening of other pathways. Perhaps this receptor plays an important role in cancer growth in lab animals but not in humans. Maybe this receptor is part of a family of closely related receptors, so interfering with one subtype of receptors has a powerful effect, while interfering with others has little effect. Perhaps the structure of the receptor varies slightly due to genetics. And since cancer cells are prone to mutation, perhaps this receptor itself actually changes over time in some patients. And maybe the receptor is right but the molecule is not. Maybe the molecule does not bind well, or is somehow being altered by the body. Maybe different people metabolize the molecule differently. Maybe the doses are not correct. But if a higher dose is given, how would that affect toxicity and side effects? Maybe the innovators should try different patients, or use this drug only in the early stages of the disease.

Many writers underscore the complexity of drug discovery (Burns 2005; Christensen et al. 2009). For example, according to Scannell et al. (2012), targets are part of complex networks leading to unpredictable effects, and biological systems have high degrees of redundancy which can blunt the effects of highly targeted drugs. These authors argue that the serial filtering using rational drug design and high throughput screening to identify highly targeted drugs is therefore limited if used alone, because the search space is enormous. There are 10^{26} to 10^{62} chemotypes that fit drug possibilities and each has many derivatives, while high throughput screening deals only with 10^3 to 10^6 chemotypes. They recommend directed iteration, even if the cycles are slow, because such iteration may be a more efficient way of searching a large multidimensional space. West and Nightingale (2009) also argue that much more attention needs to be given to uncertainty and serendipity in biological research. Bensaude-Vincent and Stengers (1996) make a similar case for medicinal chemistry.

The cancer drug illustration suggests that the scientists may have started with a narrow hypothesis that did not take into account many of the interdependencies they eventually discovered in the trial. Perhaps they did not seek to accumulate and make sense of large amounts of noisy information, they did not search far from their expectations to explore diverse consequences, and they did not keep all the possible interdependencies among the compound, cancer disease, and human biology in mind as they developed the drug. As Scannell et al. (2012) suggest happens often in pharmaceuticals, perhaps the cancer drug innovators moved prematurely into clinical trials, where the style of inquiry shifts from discovery to confirmation. So instead of staying open to possibilities, they closed in on a specific expectation. Unfortunately, the failure to confirm their expectation leaves the innovators adrift in a vast sea of possible explanations to follow up.

Abductive Learning Routines for Taking Advantage of Emergence

I propose that continually cycling through three abductive learning routines for formulating, evaluating, and reframing hypotheses about what might constitute a viable drug enables innovators to take advantage of emergence and leverage the information that is available to them. The three learning routines are not distinct steps to be executed in sequence, rather they are overlapping and entangled processes of reasoning that innovators cycle through repeatedly to materialize their new product. Together, the three routines open up the black box of 'exploration' by delineating straightforward and sensible everyday actions of collective reasoning. By extension to the rest of the subsystems in the infrastructure, the hypotheses to be formulated would relate to the problems of each of the subsystems involved. In this chapter, I outline the abductive learning routines for the project/product subsystem, and then briefly discuss why the other subsystems require similar abductive learning routines to enable drug discovery.

Formulating Hypotheses by Imagining Configurations of Interdependencies

The first abductive learning routine is using clues to imagine a configuration of interdependencies that would constitute the product, knowledge system, business model, or institutional arrangements. Not any hypothesis will do for complex innovation, which by definition means creating novel combinations of elements. Simple hypotheses about one cause leading to one effect will not work. Focusing on the drug product, innovators imagine a configuration of interactions among molecular compounds, the disease process in question, and the rest of human biology. They construct a coherent story about how a chemical compound will behave in the body against the disease. The three elements of formulating a hypothesis are the configuration, the clues, and the process of imagining. Each of these elements synthesizes useful information despite the noise, so this kind of hypothesizing captures a great deal of the available information and variation in the problem space, and makes it meaningful.

First, the content of the hypothesis is a configuration of possible interdependencies among product parts such as molecular compounds, disease processes, and the rest of human biology. The configuration conveys a great deal of information about the product. Innovators must figure out the relevant parts for a new product, but they also must figure out how these parts work together, and how they depend on each other. Focusing on the interdependencies rather than only on the parts highlights the major source of uncertainty in complex systems, where failures often arise because of unexpected interactions (Scannell et al. 2012). By focusing on a reasonable set of interdependencies among the parts, innovators attend to the possible product in action as it functions in the body against the disease. The hypothesis reflects how product elements mutually generate the desired functionality.

Second, using clues captures information, too, because clues convert existing information into directions that lead out of perplexity, as the dictionary definition suggests (*Websters*). According to Weick (2005), clues point to a world in which they are meaningful, and so give rise to speculations, conjectures, and assessments of plausibility rather than focus attention on a search among known rules to see which ones might best fit the facts.

Third, imagination 'conceives a whole design almost at once, which it then fills out and gives body to by particular association. ... The mind thinks simultaneously of specific parts and of their one organizing principle' (Engell 1981: 82–3). Imagination can be understood 'as the ability to conceive of something, seen only fragmentarily or superficially, as a complete, perfected and integral whole' (*Merriam Webster's* 1984: 415).

Evaluating the Imagined Configuration of Interdependencies by Elaborating and Narrowing

The second abductive learning routine is to evaluate the imagined configuration by elaborating and narrowing around the interdependencies. Innovators empirically inquire into the actual effects of their hypothesized configuration in order to assess the nature of the mechanisms that govern the interdependencies they imagine. In so doing, innovators surface new and deeper insights about how a chemical compound might behave in the body against the disease. Evaluating burrows into the mechanisms to explore how and why the configuration might work, what else may be going on, and what are the limits and contingencies. Innovators use the hypothesized configuration to sift through all the noisy information as they open up around possibilities to explore them, and then narrow down situated aspects of interdependencies.

Elaborating out and opening up keeps innovators poised to spot novelties and minor perturbations that may become significant, while keeping the whole in mind. Narrowing in captures the particulars of the context of the disease and produces more clues. Looking at the configuration in action assesses the nature of the mechanisms that animate the drug possibility. The innovators generate data about the usefulness of alternatives and consequences, keep possibilities open to explore alternatives, and stay actively engaged in the evaluating. The hypothesis imagines the configuration working in the context of action, and the evaluation process further contextualizes and situates the possible new product.

Reframing the Imagined Configuration of Interdependencies by Iteratively Integrating

The third abductive learning routine is reframing the hypothesized configuration by iteratively integrating across disciplines and experimental situations to accumulate and synthesize information. By reframing, the innovators holistically assess what they know so far and what they have learned. Different people see different aspects of the drug possibility and how it might function in the body against the disease. Iteratively integrating helps to overcome competency traps, push ideas, cross-check possibilities, and generate a joint representation. The innovators are not simply searching, they are actively configuring. They drop some alternatives, develop new performance parameters, and adopt new consequences that seem more promising based on their collective learning. Reframing cycles back with a new hypothesis of the configuration of interdependencies to be evaluated again.

Ansell (2011) describes evolutionary learning from Pragmatism in a way that fits nicely with my three abductive learning routines. I pay more attention to hypothesizing a configuration of interdependencies, which captures the process of innovation. Ansell (2011) pays more attention to reflexive inquiry and deliberation, which capture the political challenges of creating public solutions. A clash of different and sometimes incommensurate perspectives provides reflexive inquiry that can push development. Deliberation leads to the development of inter-subjective meaning as people probe, adjudicate, and bridge differences that can reshape prior beliefs and goals.

Together these three abductive learning routines enable complex innovation because they build on available information despite the noise, they generate new meaning, and new categories of knowledge, they keep the whole in mind, and they attend to the central unknowns in complex systems. The innovators want to find that smallest possible set of workable interdependencies that would constitute a safe and effective drug. By always looking at the interdependencies, they are more likely to use imagination and intuition. The imagined configuration that is formulated, evaluated, and reframed makes working distantly from existing knowledge easier because there is something sensible to work against.

Learning events arise from the cycling through the abductive learning routines. We define learning events as endogenous occurrences that emerge when scientists and managers learn enough about the configuration of interdependencies they are working on to indicate the next thrust of their innovation work (Dougherty et al. 2013a). These learning events capture emerging understandings in the innovation, and reflect current and anticipated knowledge resources. Learning events are moments of closure in the exploratory searching that capture enough of the whole configuration of interdependencies to enable people to see what they know so far and to identify plausible next thrusts in their innovation work. The cycles of abductive learning routines in all the subsystems produce learning events, which are intermediary models or rough drafts of the drug possibility, the redesigned strategic path, future value-creating possibilities, and possible collaborative commons. Learning events redirect the work closer to the ultimate goal of developing a good drug, a good strategic path, or a good value-creating opportunity.

Learning events capture knowledge as just described. They also coordinate work and mark progress. In our study of time pacing (Dougherty et al. 2013a), we find that scientists pace their work by the emergence of anticipated but unpredictable learning events. Pacing coordinates activities because it regulates the intensity and direction of people's attention and efforts. Normally, we think of pacing work by schedules and clocks, but I will describe how learning events can pace collective work in complex innovation systems more effectively.

Regarding using leaning events to gauge progress, participants can evaluate how 'good' learning events are, how well they define the disease or its amelioration, how quickly they are developed, or how plausible are the next steps suggested by the learning events. However, using and even arriving at learning events requires judgement (Vickers 1968). To use the learning events to gauge progress, innovators must make judgements regarding the possible facts of their discovery process and the significance of those facts. The ability to judge is enhanced if all four subsystems enact abductive learning routines to take advantage of emergence. Any new drug is a system within a system within a system, yet all the systems mutually constitute each other and are interdependent. Project scientists may judge that they have arrived at a good learning event because knowledge system scientists have already defined some possibilities, and strategic managers have projected out health care applications that can use the potential new therapy. So by eliminating some interdependencies and finding others, the project scientists may be ready to reframe their hypothesis and move towards the goal of a new drug.

The Inhibiting Influence of Existing Learning Routines Based on Simple Rationality

Unfortunately, our research also suggests that discovery scientists and their managers do not effectively implement these abductive learning routines in projects, and do not emphasize taking advantage of emergence in other subsystems. No one we interviewed said they were working abductively so perhaps they did not understand that there is an underlying style of reasoning that can structure discovery learning. Instead, scientists talked about luck and intuition, which would certainly worry managers who are responsible for spending so much money on drug discovery. Managers and other scientists also push project scientists to be more disciplined, to make better decisions, and to get clear answers. Some managers told us that the project scientists focused on science for its own sake rather than on discovering drugs. As well, we find that the other problems of innovation in the other subsystems of the infrastructure are not addressed as emergent challenges.

I suggest that other learning routines are already in place and well entrenched, so adopting the new social technologies will require breaking away from old ones. Existing learning routines rely on a simple view of rationality, and this simple rationality dominates especially the strategic and institutional subsystems of the infrastructure. Simple rationality assumes that the problem is already defined, so people need only identify their goals clearly and then identify the most effective means for achieving those goals (Vickers 1968; Joas 1996). According to Weick (2005), norms of rationality suppress abductive reasoning by dampening productive conflicts, reducing uncertainty, and driving out the active and continuous mixing of ideas. This simple view of rationality dominates the principles that are taught in schools and the approach to research in management and related disciplines (Schon 1983; Tsoukas 2005), especially in economics (Grandori 2010), where rationality is restricted to being 'logical' (i.e. based on deduction), consistent, and deductively correct.

Not surprisingly, scholars and commentators have developed alternate approaches for figuring out problems and working towards solutions for years and years, since so many of the problems society faces are emergent and undefinable. We have repeatedly developed alternatives to simple rationality, but simple rationality dominates nonetheless. Hari Tsoukas and colleagues and Karl Weick suggest three ways that simple rationality fails, each with alternatives for moving forward. I summarize these to show that taking advantage of emergence has been in the air for a long time—and maybe we need to actually move on, finally.

One way that simple rationality fails is through the 'representational' approach, which assumes a pre-given world in which action is driven by reliable prior knowledge (Tsoukas and Knudsen 2005). In a pre-given world, actors follow explicit rules to achieve their goals, and the learning process itself is black-boxed since it is assumed to happen automatically. Unfortunately, while many management researchers study processes, they black-box the processes and look only at inputs and outputs. They do not empirically observe the evolving sequences of actions and interactions that occur over time and space in actual practice (although some simulate processes in the abstract). Ironically, given the presumed rationality of this research, these scholars do not examine their unarticulated and likely incomplete understandings of the actual processes. Tsoukas and Knudsen (2005) synthesize alternate theories that highlight the importance of process versus inputs/outputs. In what they call the enactive approach, the world is assumed to be open-ended, which allows for fundamentally new and unexpected events to happen. The enactive approach highlights the personal-cum-constructed character of human knowledge, and conceives of action as experimentation, so thinking and acting are perpetually engaged in a dialogue.

Another way that simple rationality fails is through what Tsoukas (2005) calls the defensive approach to uncertainty, which presumes that noise and uncertainty are to be avoided. However, Tsoukas (2005) argues that uncertainty is valuable because it is the source of all novelty and renewal. This noisy information cannot be reduced to an algorithm or compressed to something simpler, but it can be made meaningful. Uncertainty and surprise are calls to action and challenges to invent new codes. In the process of creating meaning, people integrate elements that ordinary codes of readings do not account for, and they gain greater comprehension. Tsoukas (2005) proposes the receptive view instead, which is open to the presence of unanticipated information.

Third, simple rationality fails by abstracting away situated details and peculiarities to denote the general and universal. According to Weick (2005), a heavy reliance on analytic denotation and known rules strips away associating principles and imaginative conjectures, so when weak signals appear, conjectures tend to be conventional. In complex settings, details are essential for developing understandings because they point to novel possibilities as people manipulate their objects of study, draw on all their senses, and leverage their experiences. Being situated and contextualized draws attention to more information (Tsoukas and Dooley 2011), and allows the learner to invoke possibilities (Lave and Wenger 1991). Situated learning works in science as well, where many breakthroughs arise from focusing on applied problems (Stokes 1997; Nelson 2005), and where bedside science with patients enhances drug discovery (Gittelman 2015). For those phenomena that are inherently situated, abstraction does not generalize, it misrepresents.

I think that the three sets of social technologies developed in this book provide a way to carry out the enactive, receptive, and contextualized approaches to collective reasoning.

Insights from Science, Technology, and Innovation that Reinforce Abductive Learning Routines and Heedful Interrelating

Many scholars along with Tsoukas and Weick argue against this simple view of rationality, and offer alternative approaches to learning and organizing that encompass processes, novelty and uncertainty, and situated details. Some want to abandon rationality entirely, but others develop a richer conception of rationality that includes abductive reasoning. According to Grandori (2010: 479): 'Rationality has little to do with knowing everything, but a lot to do with following good rather than bad procedures in data gathering, hypothesis testing, assessing probabilities, and comparing options.' She summarizes others who argue that rationality includes systematic, valid, and sound methods for constructing the knowledge on which decisions are based—which is exactly what the abductive learning routines do for complex innovation.

I summarize three different literatures relevant to complexity that arrive at processes that are very similar to the empirically grounded abductive learning routines that we have developed (Dunne and Dougherty 2016), and to the heedful interrelating developed by Weick and colleagues. I do not claim that all these scholars confirm our ideas about abductive learning routines, since these various ideas have been developed for different purposes or for different dimensions of learning. However, I do claim that very similar ideas have been around for a very long time. In fact, the depth and strength of these ideas is surprising. Organization theorists have been exploring the role of the alternate approaches to learning like abductive learning in complex domains for decades, sometimes directly by using the idea of abduction, and other times indirectly by describing learning processes that mirror descriptions of abduction by Peirce and others. However, there is little cross-citation, suggesting that people (including myself) continually develop similar ideas from their own view rather than try to build a more coherent theoretical understanding. This 'conceptual churning in place' indicates that ideas about learning in complex domains are still immature, ad hoc, and idiosyncratic. I hope that this book will help move us beyond conceptual churning to applying these ideas to serious social problems.

Abductive learning routines and heedful interrelating fit well with: (1) processes of doing science as described by scientists themselves as well as by sociologists of science; (2) studies of complex problems, professional work, and decision-making; and (3) with organizing for innovation.

Science and Abductive Learning Routines

The conventional view of science draws on the Mertonian stereotype of the scientific method that reflects the simple view of rationality. This stereotype relies on common values and norms for validating claims and publishing results. This approach to science, according to Schon (1983), Garud et al. (2011b), and Tsoukas and Knudsen (2005), is to follow a linear, step-by-step sequence from basic to applied problems, relying on the logic of deduction. Grinnell (2009: 4), a practising microbiologist, explains this linear approach to science as follows:

> According to the linear model, the path from hypothesis to discovery follows a direct line guided by objectivity and logic. Facts about the world are there waiting to be observed and collected. The scientific method is used to make discoveries. Researchers are dispassionate and objective.

This linear, dispassionate model of science is assumed by many to be the only way that science can be done. If people do not follow these procedures, they are not doing science.

While this approach to learning may fit problems of 'normal science' that incrementally exploits established knowledge, it cannot enable the learning for innovation in infrastructures of complex innovation systems. Scientists themselves point out that real science does not adhere to this linear model. According to Grinnell (2009: 4):

> I believe the linear model corresponds to a mythical account—or at least a significant distortion—of everyday practice. Rather than linear, the path to discovery in everyday practice is ambiguous and convoluted with lots of dead ends. Success requires converting those dead ends into new, exciting starts.

Others also argue that learning under complexity is not about deducing specific conclusions from a closed set of general premises, because existing understandings are too weak to provide useful predictions, and any results are ambiguous (Nightingale 2004; Pavitt 1999; Nelson 2005). Deduction in such situations cannot generate useful knowledge, because when the number of possible explanations is very large people cannot feasibly isolate and select particular options.

Grinnell is not an outlier. In his address to the American Institute of Biological Sciences, on the occasion of receiving the distinguished service award (and on his 96th birthday), Ernst Mayr (2000: 896) explained:

> The basic philosophy of biology, as it developed in the last 50 years, has become quite different from the classical philosophy of science as it prevailed from the Vienna School of Carnap and Neurath to Popper and Kuhn. In the rejection on the one hand of all vitalistic theories and of such concepts as essentialism, determinism, and reductionism, and their replacement by an acceptance of the frequency of random events, plural solutions, the importance of historical narratives, multiple causations, population thinking, and the greater importance of concepts than of laws in theory formations, the new biology has undergone a complete revolution.

Mayr went on to say in his speech that our understanding of the basic biological phenomena is 'remarkably well advanced'. But '(w)here our knowledge lags behind is in the understanding of complex systems'. He listed three: the developmental system (from a zygote from the fertilized egg to the finished adult), the central nervous system (will we ever know about the interactions of the three billion neurons of our central nervous system, each with up to a thousand connections with other neurons?), and the ecosystem. These are basic research questions in biology, but their ongoing exploration will also inform drug discovery in a variety of ways.

Those who study scientists also draw very different conclusions about how real science works. Nightingale (2004) and Schon (1983) argue that scientists and professional practitioners use the discovery style of inquiry to take advantage of emergence, one that is based on active experimental intervention. Rather than isolate a particular option, knowledge professionals create something new to learn from by intervening to test specific mechanisms. These interventions build up understandings and inform judgements about what might be working (Hacking 1983). With this discovery style of inquiry, innovators do not ask if something works, but rather they ask how something works.

As outlined in Chapter 1, many scholars of science and technology point out that scientists, engineers, and other knowledge workers figure out complex problems even though existing theory is too weak to suggest solutions. Latour (1987) highlights the emergent nature of scientific knowing when science is 'in the making', and analyses science in action as a process through which inventions and discoveries become accepted. The knowledge of science provides some initial understanding of a complex landscape, and scientists use what they know to get a 'glimpse of the possible' (Fleming and Sorenson 2004). In her ethnography of academic microbiologists, Knorr Cetina (1999: 92) finds that the scientists did not try to understand the numerous problems that arose in their experiments because 'their attempts to understand a living organism, of which little is known, quickly reached its limits'. Instead they would treat problems by 'varying components of the experimental strategy until things worked out, not by launching an investigation of the cause of the problem'. Schon (1983) also argues that situations of professional practice are complex, uncertain, unstable, and unique.

Learning for Complexity and Radical Innovation and Abductive Learning Routines

A number of other scholars address learning under complexity and radical innovation. I summarize some of these ideas for each of the three abductive learning routines, to suggest additional ways that the routines can be enacted in practice.

FORMULATING HYPOTHESES BY USING CLUES TO IMAGINE A CONFIGURATION OF INTERDEPENDENCIES

The literature on complex learning emphasizes constructing the problem itself by creating wholes and connections, similar to our first abductive learning routine of formulating a hypothesis about a configuration

43

of interdependencies. Dunne and I build on Weick's (2005: 433) ideas about imagining a reality in which a tangible clue is meaningful:

> Current use of this broadened sense of abductive reasoning is found in the work of people such as Ginzburg (1988), Harrowitz (1988), and Patriotta (2004) who argue that the conjectural paradigm, grounded in abductive reasoning, is the foundation of inquiry. The basic idea is that when people imagine reality, they start with some tangible clue and then discover or invent a world in which that clue is meaningful. Imagination 'conceives a whole design almost at once, which it then fills out and gives body to by particular association.... The mind thinks simultaneously of specific parts and of their one organizing principle' (Engell 1981: 82–3). This act of invention is an act of divination that has a close resemblance to detective stories.

Weick (2005) emphasizes 'a whole design'. Imagination can be understood 'as the ability to conceive of something, seen only fragmentarily or superficially, as a complete, perfected and integral whole' (citing *Merriam Webster's* 1984: 415). 'Imagination is the power to present in concrete, particular forms and expressions what before had been only general and abstract knowledge, hazy feeling, or impression' (Engell 1981: 101).

Others also talk about combinations, similar to our idea of configurations. For example, Fleming and Sorenson (2004) suggest that scientists in complex domains formulate useful new combinations among interdependent elements. Schon (1983) details the process of formulating a combination with 'problem setting'. He argues that when situations are complex, unique, and fraught with value-conflict, ends are not fixed and clear but rather are confused and conflicting, and there is no problem to be solved. Professional practitioners must first construct the problem from the materials of problematic situations which are puzzling, troubling, and uncertain.

Through the process of framing the problematic situation, people organize and clarify the decisions to be made, the ends to be achieved, and the means which may be chosen. Problem setting is the process by which innovators select what they will treat as 'things' of the situation, set the boundaries of their attention to it, and impose upon it a coherence which allows them to say what is wrong and the directions in which the situation needs to be changed. It is a process in which, interactively, innovators name the things to which they will attend and frame the context in which they will attend to them. Similarly, Pragmatists like Dewey say that inquiry is a progressive determination of the problem (Ansell 2011).

Additional ideas for formulating a hypothesis as a configuration come from Gavetti and Levinthal (2000), who say that 'cognitive representations' of the problem are central to discovery, because they seed and constrain the processes of experiential learning. Even crude representations provide a powerful starting point, constrain the process from wandering into less attractive places or getting stuck in local places, and allow for a broader examination. Focusing on defining a decision problem, Grandori (2010) says that the problem should not be formulated as a gap in expected performance, because then people would tend to categorize the problem as one of a known type and apply solutions that are supposed to work. Nothing new is developed. Instead, the problem should be formulated as a performance potential, with causes understood as available resources or alternatives (Penrose 1959), and effects understood as useful consequences. The resources should have multiple potential, and are hypothesized to be put to certain uses which would produce valuable consequences. Citing Hanson (1958), Grandori says that this is theoretical abduction, which refers to formulating theory-based, causal hypotheses from which the observed or sought action/consequence chain would follow.

Studies of radical innovation come to similar understandings of formulating an initial understanding. According to Van de Ven et al. (1999), the beginning of a radical new product is an expanding, divergent vision and process, where people discover what courses of action are possible with a new product idea, what outcome goals and criteria they prefer, and in what kind of institutional contexts they will work. Because the product is complex, innovators cannot assume that familiar regulations and market dynamics will apply. Instead, they must gain experiences with alternate ways to think about a new technology, the various components and architectures that are possible, and different testing and evaluation procedures. Van de Ven et al. (1999) call this process 'learning by discovery'. The authors discuss the need to combine elements for the product that identify the kinds of things the product will do for whom in what market and institutional context. Leifer et al. (2000) also argue for synthesizing new and non-obvious insights from bits of disparate technical information, using a loose, associative thinking process, not a sequential process. This loose process is necessary to maintain uncertainty and avoid premature closure. Leifer et al (2000) also say that innovators need to come up with a number of technology development paths, should any single one not bear fruit.

EVALUATING HYPOTHESES BY ELABORATING AND NARROWING

The second abductive learning routine involves evaluating the problem that is set by elaborating and narrowing around the interdependencies, to see how this framing works and to improve it. The literature reflects our ideas about experimenting in a tinkering mode to shape the situation and to probe it further. Fleming and Sorenson (2004) point out that scientists engage in trial and error learning because they do not have sufficient knowledge to predict all the interactions that might occur. I repeat Paavola et al. (2006)'s quote of Ernst Mayr on Darwin from Chapter 1, because it exemplifies going back and forth:

> [Darwin's] procedure does not fit well into the classical prescriptions of the philosophy of science, because it consists of continually going back and forth between making observations, posing questions, establishing hypotheses or models, and testing them by making further observations, and so forth.

Evaluating asks the big question of how this configuration works in the context of use, and for new drugs, how it works in the body against the disease. Additional questions for evaluating would focus in on particular interdependencies to figure out the connections that do and do not matter.

Nightingale (2004) and Schon (1983) suggest that evaluating is a process of opening up to explore rather than closing in to confirm. As they tinker with possibilities, Nightingale (2004) explains that scientists would use a variety of technologies and methods to select explanations for their problem, and test implications against each other. Scientists would reject those that are not better than prevailing theory, but they still would have a lot of possibilities. Nightingale (2004) says that scientists become reflective practitioners. Schon (1983) elaborates on this process. Practitioners use reflection-in-practice to experience surprise and puzzlement and see the unexpected, so that they can reflect on prior understandings which have been implicit in their behaviour. Practitioners surface and criticize their initial understanding of the phenomenon that now seems surprising or unstable, construct a new description of it, and test the new description by an on the spot experiment. Reflective practitioners step into a problem situation, impose a frame on it, and follow the implications of the disruption thus established while remaining open to the situation's back talk.

In Denrell et al.'s (2004) model of complex problems, problem solvers evaluate the mental model they construct of their problem situation by making predictions, and incrementally using information generated by the deviations from the predictions to update and rework their model. Grandori (2010) argues that the logic of discovery relies on systematic, hypothesis-driven data gathering. Decision-makers should discard hypothesized effects that are not observed, and include unexpected effects that are observed. They should not fix evaluation parameters ex ante, but empirically inquire into the actual effects, and compare alternatives on their relative capacity to produce unspecified but rankable streams of consequences.

In their study of radical innovation, Van de Ven et al. (1999: 184) find that innovators engage in a 'nonlinear cycle of divergent and convergent activities that may repeat over time and at different organizational levels if resources are obtained to renew the cycle'. These findings are similar to our finding of elaborating and narrowing, although Van de Ven et al. (1999) refer to big cycles of activities while we refer to more immediate probing and learning. Their divergent process explores the broad vision, while their convergent process chooses among these elements with trial and error learning. In the convergent process, innovators narrow the innovation by exploiting a given direction and leveraging a given set of relations in particular institutions. This convergent cycle channels subsequent divergent behaviour, because once innovators narrow their attention to a particular technical design or market niche, whole new sets of issues open up.

REFRAMING BY ITERATIVELY INTEGRATING ACROSS DISCIPLINES AND SITUATIONS

The third abductive learning routine is to reframe their problem setting or hypothesis, by adapting the initial frame. Scholars highlight a variety of dynamics that reframe the problem or hypothesis. For Schon (1983), reframing helps practitioners restructure a situation, so that eventually they can say that the theory fits the situation. Practitioners are in conversation with the situation so that their own models are shaped by the situation. They use iterative tinkering to understand the situation by trying to change it, and consider resulting changes not as a defect of the experimental method but as the essence of its success. Grinnell (2009) discusses Nobel Laureate Max Delbruck's principle of limited sloppiness. Sloppiness is used in the sense that our conceptual understanding of a system under investigation is frequently a little muddy. Consequently, experimental design sometimes tests unplanned questions, as well as those explicitly thought to be under consideration. Unexpected results can emerge and lead to more important findings. Scannell et al. (2012) quote Sir James Black on 'obiquity', the art of looking for one thing and finding something else. Gavetti and Levinthal (2000) emphasize changing the cognitive representation to create a fresh perspective and to see new aspects of the problem. Grandori (2010) suggests that decision-makers reformulate their problem definition by creating new performance objectives that reflect new alternatives and consequences, and avoid simply lowering aspiration levels.

To summarize, scholars discuss a variety of dynamics for learning under complexity, but many fit with our three abductive learning routines to form a reasoned approach to problem setting and solving. Taking advantage of emergence relies on science and on proceeding from reason. Innovators formulate their hypotheses by imagining a whole world, problem setting, constructing a representation of the situation, and defining the problem as a potential for accomplishing useful consequences. Innovators evaluate their hypotheses by tinkering and working back and forth to surface and critique what they are thinking, and to push themselves to find other better ideas. Innovators reframe their hypotheses based on new insights about what they think they are doing, and on unexpected results that suggest new performance objectives.

Organizing for Innovation and the Abductive Learning Routines

A third category of research is also consistent with the organizing side of the three abductive learning routines, and highlights the need to organize the collective process of learning for complex innovation. Learning routines are social, in that a routine is a recognizable, repetitive pattern of interdependent action that involves a variety of people who collectively engage in the pattern. The infrastructures for systems of complex innovation depend on thousands of people working together.

First, many quantitative studies demonstrate that organizing for something like heedful interrelating enhances the ability to innovate. While they do not specifically use the concept of heedfulness, several studies find that a heedful-like organizational context fully mediates the relations between inputs like new technologies, market intelligence, and other resources and outputs like more profitability, more new products, and more technologies. In other words, the inputs do not lead to the outputs unless heedful organizing processes are in place. For example, Siren et al. (2012) study strategic learning, which involves disseminating strategically important information, interpreting new information openly, critically reflecting on it and questioning biases, and implementing new ideas in products, processes, and organizational procedures. They conclude that to take advantage of technologies and market information that are gained from entrepreneurial activity, organizations need a strategic learning capability.

Gibson and Birkenshaw (2004) find that an organizational context characterized by stretch, discipline, support, and trust mediates the relationship between the external competitive context and organizational performance. Foss et al. (2011) find that practices of intensive vertical and horizontal communication, rewarding employees for sharing and acquiring knowledge, and high levels of delegation mediate the effective use of customer information—that is to say, without this kind of organizational context, the customer information will not be used well. The abilities to work with and learn from other organizations also supports innovation. For example, Rosenkopf and Nerker (2001) find that when organizations build on the learning of other organizations, they develop more impactful technologies. Rothaermel and Alexandre (2009) find that ambidexterity in technology sourcing enhances firm performance.

Qualitative research identifies organizing processes that also reflect heedful interrelating. Van de Ven et al. (1999) studied multiple cases of radical innovation, several at 3M Corporation. They find that an organizational capacity for gestation is essential to the formulation of radical new products. Gestation opens the organization to the importance of chance. People work independently to develop new ideas, and these ideas intersect by chance with independent actions of others. The intersections provide occasions for people to recognize and access new opportunities and potential resources. When these occasions are exploited, people adapt and modify their independent courses of action into convergent, interdependent actions to mobilize efforts to initiate an innovation.

In a more recent analysis of 3M innovation, Garud et al. (2011b) find that people use their own initiative to take advantage of opportune moments in multiple ways. First, the organization maintains long-term technology development processes with technology platforms that everyone can draw on for a particular opportunity. Intermediary insights and artefacts become integral parts of technology platforms, and are combined and recombined over time. Second, the organization fosters ongoing interactions among employees to use the company's diverse resources via multiple 'agentic orientations', or a variety of ways that people can take agency to work with different types of complexity. Third, there are simple rules, like employees can spend 15 per cent of their time on their own ideas. Finally, innovation narratives link past, present, and future, enabling people to manage the tensions among divergent complexity forces.

My own work with colleagues (Dougherty et al. 2005) develops complementary ideas for an organization-wide competence for discovery. We build on Giddens's (1979) theory of social structure as the combination of social rules and resources, to define the rules and resources that support innovation. I think that 'agentic orientations', individual initiatives, and abilities to collaborate are enabled by three social rules that guide everyday behaviour: share responsibility for the entire innovation project, so everyone contributes throughout; all knowledge is valuable, so respect expertise and make your own expertise accessible to others; and look for opportunities to add value by continually exploring options and alternatives to problems. These social rules operate along with social resources: access to time and attention of other people; access to knowledge; and access to a variety of options to tackle problems. These rules and resources are fairly straightforward ways to make innovation the default option for organizing. The non-innovative rules reflect the limited view of rationality and hierarchical organizing: separate responsibility so people can be held accountable; focus on outcomes to control how your own expertise is applied; and seek to eliminate problems to reduce reliance on others.

I conclude from these studies along with my own research that innovators, managers, regulators, and other participants in complex innovation systems need to organize not only projects but also their organization and their inter-organizational relations to take advantage of emergence. The whole infrastructure needs to be ready to grab opportune moments. Even if an organization like 3M is ready to take advantage of emergence, its efforts would be stymied continually by the need to stop and create new standards, new business models, or new collaborative arrangements for each new product.

Conclusion

I have only scratched the surface of ideas in management that fit with something like abductive learning routines, and emphasize emergence. My thinking has been informed by many others such as Theresa Lant, Marlena Fiol, Anne Smith, and Jim March to mention just a few, who are not cited, in part because I have absorbed their ideas into my own. However, emergence is the norm of everyday life, according to Weick (1995), so the real danger is to not enable emergence. Most of us are used to emergence in our daily lives, while for strategic managers, competitive advantage is often only temporary so they need to focus on continually generating new sources of advantage, all of which emerge (Brown and Eisenhardt 1998).

I do not claim that all these other ideas confirm my ideas about the abductive learning routines. But I suggest that abductive learning routines synthesize aspects of many of these insights about learning and innovating into a new process that allows people to take advantage of emergence. My contribution is to develop in more detail how people can figure out how to create new products and services that resolve important problems like health, economic development, or ecological damage despite the noise and fragmentation. I will lay out in much more detail exactly how people can work together in and across the four subsystems with alternate ways of thinking and doing that are none-theless sensible. The abductive learning routines are clear, observable, practical approaches for figuring out complex problems, where answers do not exist.

Participants in the pharmaceutical infrastructure may be able to break out of prevailing social technologies that inhibit abductive reasoning because so much money is at stake. I see in the discussions and reviews in this industry sincere efforts to collaborate more fully and to work on project and knowledge system problems more coherently. I think that transforming the entire infrastructure is the only way to generate poten-tial profits, and, more importantly, improved health care. In the next four chapters, I focus on each of the subsystems. Each chapter begins with a brief synopsis of the transformations in management for each kind of problem during the past forty years or so that have created breaches in the walls of hierarchy, decomposition, simple rationality, and command and control management. Then I detail the next wave of transformations that complexity introduces, and describe how people can enact these qualitatively new social technologies. I hope to move beyond breaking the walls down and get on with building new ways for collective thinking and doing for complex innovation systems.

3

The Project Subsystem in the Infrastructure for Complex Innovation Systems

Navigating in the Labyrinth

The project subsystem addresses the most central challenge in complex innovation systems—creating, developing, and materializing a new product or service that actually addresses a particular aspect of the broad objective of the complex innovation system. In the case of pharmaceuticals, the broad objective is to generate drug therapies to address unmet medical needs. A project team ideally produces a particular drug that behaves well in the human body against a particular disease process. In other complex innovation systems the projects also solve particular problems. In alternate energy, project innovation might produce wind generators that not only work off-shore better than existing windmills, but also address particular problems with navigation or bird migration. In education, project innovations might create and implement a particular classroom approach that enables teachers to enhance learning with diverse students, including non-native language speakers.

We adopted Denrell et al.'s (2004) metaphor of navigating in a labyrinth to characterize the project subsystem's working processes because of the complexity involved (Dunne and Dougherty 2016). In non-complex problems, innovators have alternates to try out, each with some known probability of success. Denrell et al. (2004) explain that this kind of challenge is like a T maze, where one picks from one of two choices, each with some probability of success. However, complex problems do not involve known alternates or established probabilities. Innovators do not know how current activities relate to the end objective, because they do not know what choices are involved or how each choice might interact with the substance of all other choices. Rather than move linearly through a simple maze, innovators working on complex problems continually confront surprising barriers and new options, and must navigate their way through a labyrinth by making choices based on some intermediary model of their situation, probing and learning about that model, and revising the model as they proceed.

Even though they are navigating in a labyrinth, product and service innovators follow a disciplined development process. Innovators bring an idea into existence, or materialize it in such a way that it functions in the real world. The project subsystem in the infrastructure for all innovations, including complex innovation systems, builds and deploys these functioning, concrete solutions for real problems. Managers cannot simply devise a new idea and hand that to people to 'go do it'. I emphasize this central development process in the project subsystem because it is so often overlooked. I am sure many of us receive emails at work regarding how we (meaning you and me) ought to do something about X: improved career development for our students, better mentoring programmes for our new hires, new market possibilities, and so on. In fact, the idea generators ought to execute a new product or service development process in which a team of experts identifies all the elements that are needed in the new programme, figures out how these elements work together to produce the desired functionality, works with operating units that are involved to flesh out details, fits the new programme into the rest of the organization, and actually builds the programme so that it works by implementing and revising early efforts to work out all the bugs.

Ignorance about the project subsystem has a long history, and I see many instances in the news that confuse ideas with actual working programmes for many social programmes (health, education, policing). While at present many firms in many industries generate ongoing revenues by generating a stream of new products over time, active innovation is a relatively recent transformation. The social technologies that enabled the rise of large enterprises in the nineteenth century focused on how to optimize existing assets in production or in processes for steel making, automobiles, chemistry, and so on. These organizations generated ongoing revenues by scaling up and speeding up existing steps in the process, not by creating new products or services. If we fast forward to the 1970s, we see that firms in other countries like Japan and Korea eclipsed US and European firms in automobiles, steel, electronics, computer chips, and other areas. They did so by innovating products, processes, and strategies that met emerging market needs and leveraged the potential of technologies and sciences in new ways.

Developing a well-functioning project subsystem for innovation transformed hierarchical managing and organizing by prompting a new way to figure out how to create products, and a new way to coordinate across functions. The first breach in the walls of hierarchy came from studies in the 1970s showing that the primary cause for the failure of new technology-based products was the failure to meet customer needs. While not so surprising, organizations managed technologies and markets in separate functions. For new technologies, R&D would push out products, but it took a very long time to end up with something that generated revenues. I recall a case about a new rubber-like material developed at DuPont. Innovators spent years trying out the new material in various markets and redeveloping it to improve functionality, but nothing ever came of the material. I no doubt have the story wrong, but I clearly remember the emphasis on technology for its own sake, and that the problem with innovation came from customers' failure to appreciate this great new stuff. Product innovation was considered to be very risky, but the failure to carry out innovations correctly generated the risk.

Studies in the 1980s (Souder 1987; see Cooper 1998 for detailed history) showed that multi-functional teams generated more successful new products. During this time, academics and practitioners developed new ways to create, design, and develop innovations that made the process much less risky. Phase review procedures detailed steps such as early product concept development, market and technology feasibility testing, complete development, manufacturing, market launch, and ongoing product management. Each phase encompassed specific steps and procedures to be carried out and the necessary deliverables to transition to the next phase. Related developments fleshed out the need for an early and full product concept development (Bacon et al. 1994), and concurrent engineering to manage the overlap between designing and manufacturing. These processes cut laterally across the functional hierarchy, and depended on multi-functional teams throughout the development cycle. Innovative companies now understand that innovation is a team sport (Tushman and O'Reilly 1997).

A second transformation addressed coordinating this lateral flow of work across the hierarchy. Clark and Fujimoto (1991) studied worldwide automobile production in part to figure out why Japanese companies could go from initial design to market launch in about thirty-six months, while it took US companies sixty months and European companies even longer. They found that the product design, manufacturing, and marketing activities proceeded in parallel in Japan, while these functions worked separately to complete all their specific tasks in the

US and Europe. Each function handed the product off to the next function by transferring codified instructions that left out all the tacit information. US firms used this sequential process to reduce costs and to assure that everything was correct. The Japanese parallel process required that manufacturing begin to develop the tooling before engineering finished the new car design, and so incurred some potential costs if designs required different manufacturing. However, parallel development significantly reduced overall costs by reducing the endless engineering changes generated by the sequential process, and the additional marketing costs that arise from trying to sell a new product five years after people had assessed market interest.

In explaining how these functions coordinated their joint work, Clark and Fujimoto (1991) provide a nice analysis of heedful interrelating in practice. The authors highlight anticipation and appreciation. Each function anticipates the needs and problems in next function, and they appreciate the others' special needs. For example, manufacturing is constrained by costs and cannot invent an entire new kind of factory for each new car, while marketing needs time to develop market interest in new functionalities. Engineering needs to appreciate these problems, while the others need to appreciate engineering's challenges. Anticipation and appreciation enable all the functions to reduce costs and devise workable designs. The sequential American and European approach to coordination became known as 'over the walling', whereby each function would throw its part of the work over the wall to the next. One marketer at another firm recalled this time of over-the-walling. He said that they knew their designs or customer assessments were flawed (because they could not be done well in isolation) but they threw them over the wall anyway and then would 'run and hide'.

Complexity adds new wrinkles to figuring out how to design and develop the product, and to coordinating the work. Both these processes for project innovation must transform again to accommodate complex products. Complex products cannot begin with a clear product concept that encompasses most of the issues involved because the innovators cannot know what all the elements will be or how they will interact. Innovators cannot proceed in a linear fashion through predefined stages and gates, since possibilities will emerge suddenly and unpredictably. With regard to coordination, most auto manufacturers have now reduced development time to twenty-two months. Innovators can coordinate across functions using clock-time to regulate the intensity and direction of people's efforts and attention. Since each phase is at least somewhat predictable, innovators pace their work by the passage of clock-time (Brown and Eisenhardt 1997). But complex innovations take considerable time: thirteen years now for new drugs, and eight or more years for other radical products (Leifer et al. 2000). Ineffective development processes may produce some of this extended time, but complexity does as well, since activities cannot be predicted: unexpected alternatives arise and need to be explored, experiments need to play out, and people learn only with numerous trials of possibilities.

In this chapter, I first illustrate the complexity of projects in drug discovery and why innovators need to take advantage of emergence. Then I illustrate how the three abductive learning routines provide the new process for figuring out how to design and develop complex new products, and how event-time pacing coordinates efforts over such a long product cycle time. I conclude with the significant challenges that can inhibit these two new processes for innovation project management unless the other subsystems also use abductive learning routines to address their own problems.

Project Complexity and Taking Advantage of Emergence

Some colleagues and many peer reviewers find it difficult to believe that pharmaceuticals involves complexity, because advances in biotechnology enable people to screen millions of compounds against a possible target protein, using what the industry itself calls 'industrialization'. Recombinant DNA techniques enable the mass production of proteins that can be processed robotically using thousands of plates that each hold about 1,300 samples, and many academics have extolled the breakthroughs from all this screening (to be discussed in detail in Chapter 4). However, as one scientist told us, what if you find that 300,000 compounds 'hit' or bind to the protein? Now what? The problem remains complex despite, or as Scannell et al. (2012) suggest, because of the massive screening.

We worked our way up to the idea of abductive learning routines from our interviews. A number of the scientists we interviewed talked about using intuition and luck. At first we thought they were not thinking carefully and deliberately, but then realized that they were talking about how they deal with emergence. Intuition refers to using considerable domain experience and expertise to leap to a holistic understanding about a phenomenon, based not on rational analytics but on judgement, implicit learning, feelings, and incubation (Dane and Pratt 2007). Intuition enables people to take advantage of emergence, because it allows them to use imagination to leap to new worlds (Weick 2005).

Luck also enables people to take advantage of emergence by spotting the serendipitous manifestations of signals that might lead to a drug.

For example, a computational chemist explains how they work imaginatively and creatively, not by blasting things to death:

> We are dealing with something called sparse data. We don't have the complete picture, and we don't take every molecule and blast it to death, blast it against every acid based compound, for example. That is not the way it works. We have this big universe and we are effectively shooting at different points because we feel that that is where the drugs are…It is a fusion of many ideas that come from many people today with different perspectives using data generated. Computational chemistry is one of them, or high frequent screening is another that works completely opposite the therapy scientists, or statistical analysis of large data bases. There are many different ways to get there.

Because of the sparse data, they work towards where they 'feel' the drugs are in the vast chemical and biological spaces of human diseases. He also points out the collective and heedful nature of their searching, and how they fuse insights from many sources of different expertise. This computational chemist also explains that they must organize the information they generate, categorize it properly, and 'present it to our users our scientists, in a way that would allow them to make decisions, what to do next and what molecules to design, what compounds to advance'.

This next comment also emphasizes the tacitness of their knowledge and the importance of experience:

> A lot of the companies don't have formulation expertise. We have a whole department here…so we bring them in and sit everybody down and just start talking about it. A lot of expertise is not in papers, it's not in patents, it's just experience of working with difficult molecules on the bench.

He begins with his expertise in chemical formulation or preparing the drug to be used. But he points out how they rely on each other to talk things out, and perhaps leverage their collective abilities for intuition about working with difficult molecules.

The mention of clues also got us towards developing abductive learning. According to the dictionary, clues are something that lead out of perplexity, or a fact or object that helps to solve a mystery or a problem. Clues crystallize the noisy and fragmented information into directions out of perplexity. Scientists talked about using clues to new molecules from the molecules that they know about that may have some features similar to what they are developing for the new disease. For example, using screens, one said: 'we find a few compounds that hit the target and are doing what they are supposed to do'. This is a plausibility clue that works in a way they hypothesized about. Another biology director talked about specialists who look at the compounds in partnership with the project team, and study the chemistry. Rather than come up with a clear answer, the specialists say they are interested in these compounds. They say: 'it looks like I can work with this compound, but this other one is not good because it looks like detergent'. This example shows a similarity clue that specialists use to weed out certain compound types because they 'look like' a detergent (which binds to everything). The director explains: 'You get to know these compounds very well. We can actually look at them and think that looks nice. I can work with that.' 'Looking nice' is an intuitive perspective about a clue that they use to hypothesize a workable chemical compound. They use compounds that they 'find attractive', and that they can work with.

Project Subsystem Abductive Learning Routines

Intuition, luck, and clues all lead to the idea that the scientists are formulating hypotheses about what might work as a drug from the beginning of their discovery efforts. These are abductive hypotheses since the scientists cannot deduce from a complete theory. We also realized that these hypotheses concerned configurations of interactions between the chemical compound, the disease, and the rest of human biology. Even though the ideas just mentioned concern a chemical compound, they 'look nice' because of their interdependencies with human biology. A director of chemistry at another firm said that when they are looking for a 'lead' compound (one they will settle on and develop in detail) by exploring binding properties, they also explore all the other properties that would make it into a drug, including its absorption, distribution, metabolism, excretion, and toxicology.

A director at a third large firm explains how they rely on many teams of experts to help formulate their abductive hypothesis about a configuration of interdependencies:

> We create teams of scientists that are working at various phases of the discovery process . . . there is a team working on discovering genes of interest, . . . there is a team working on validating those genes, . . . there is a team working on converting these genes into an understanding of a protein at a level that enables us to begin to approach it from a drug discovery process, . . . to design drugs against that target. Then you have a team that gets involved with crafting that drug that modulates that target . . . By forming teams you are creating synergies around the expertise that is required as well as following a more robust triage process.

They create synergies around the expertise that generate a drug possibility. The hypothesis seems to slowly emerge, but we think that in fact they cycle repeatedly through all three abductive learning routines to arrive at the lead compound, and that represents a learning event. Then they cycle again to hone the compound's properties against the disease. This comment also suggests heedful interrelating around collective expertise and intuition.

Chemists heedfully work with the biologists to develop targets and molecules, as this director of biology suggests:

> Many of the chemists stay very much in tune with the biology so they are aware of good targets, why they are good targets, and many suggest ideas for what are good targets. They follow the chemistry literature extensively so they know there are molecules out there that modulate the targets we are interested in, or targets that are related to the ones we are interested in, so they can help us get to the starting off point to get molecules that we think may lead us toward identifying a drug. They help us with the whole strategy for how we are ultimately going to come up with a drug at the end of the day.

Staying in tune reflects part of what enables order, like knowledge, to emerge according to Chiles et al. (2004): agents purposefully pursuing individual plans based on local knowledge, and continuously adapting to feedback about the actions of others. The scientists seem to see a whole drug possibility emerging from their different activities and perspectives. Together they build these ideas into a starting-off point that configures good targets and molecules for addressing a disease.

Another chemist talked about wanting to understand how the molecule behaves in the body against the disease. This behaviour implies a story or narrative about the possible drug that conveys ongoing action of the product in the context of use.

Project innovators also evaluate their hypothesized configuration by elaborating and narrowing around it. This genomics expert describes narrowing in on the target component of the hypothesized configuration, and then elaborating out around it in many dimensions to explore other details in the disease—different people may have different genetic makeups that respond differently to a drug, or the target that is expressed is similar to other proteins that may also bind to the drug:

> If you think of a target and all the things that are important in that from a genomic perspective you need to know the variance of that target in human beings, are there splash variances, are there sniff variances or gene variances that might cause a difference in response to your drug and all that. If you want the drug to be potent and selective and not hit off target, you need to know all the close family members that you have to avoid so . . . you have to

look at the family tree and work out your best selectivity paradigms and things like that. You need to know where your gene is expressed in human tissues to make sure you have a heads-up if it is expressed somewhere it might cause a problem down the road. That is something you ought to check out and make sure you know about at least.

Evaluating depends on elaborating across models and experimental settings, and asking good questions, as this chemist explains:

> So you test your molecule and you get some results back and it is too good to be true. And if it is cell based now you go to a tissue based test and see if it is good there, and if it is and it works for an animal model... You can go to several of these models and see how really robust your model is.

They look across experimental conditions to assess how robust their model really is, but they need to judge those possibilities based on concrete, hands-on explorations.

Another scientist explains evaluating further down in the development process to assess the formulation of the drug to make sure it is stable. He stresses getting a 'total answer' about the configuration, since even at this stage complexity lurks:

> You have to address the total answer. [For example] I put my product out for long term stability testing... Somebody without discussing it much might say this is good but here I think there is something there... OK then I would like to say what is this new condition, can you tell me the structure? Then I would like to know why did this structure emerge only in this condition and why not in others? Can we look at the kinetic profile and do some prognosis and make a conclusion on whether it was the chemistry of the drug or the way we process the product...

He describes many clues, diverse tracks, and heedfully interacting to develop a sense for the total answer.

Evaluating is a 'stepwise process' as this scientist explains:

> If you know that a 100 mg dose gives you 50% inhibition of this marker in the skin and you know from your mice studies that the 50% inhibition of a skin marker leads to efficacy in a mouse tumor, then you will try to prove that that dose that produces a 50% efficacy will be efficacious in humans. It is a stepwise process that you build the strength of confidence that it is going to work.

He describes a series of clues that they put together to build confidence, and to identify next experiments and next questions they need to explore. They do not arrive at simple, clear answers and so must take advantage of emergence to pull bits of ideas together around their hypothesis. Another chemist emphasizes the empirical process as well, because 'you cannot predict outcomes efficiently'. He explains that they need to experiment with different models, observe, and 'take our best guesses' to move molecules forward.

The third abductive learning routine is reframing by iteratively integrating across expertise and experimental situations. I have illustrated these iterative processes already. But reframing also involves deliberation that examines underlying assumptions, surfaces different views, and develops ways to bridge differences by rethinking the hypothesized configuration of interdependencies. This scientist describes changes in the industry that now include more reframing:

> Many years ago the industry would work on a target, find a drug, and throw it into the clinic and go work on another target. Now we stay with a target longer, and we make our first drug and we know it probably is not going to succeed and as it is tripped up we use the learning from that to recycle and try and make an improved molecule.

They used to work step by step, but now they stay with a hypothesis longer, and cycle more often around that possibility to rethink and improve the molecule.

Reframing occurs across groups as well. A formulation chemist explains that sometimes the upstream group sends a molecule to formulation that these chemists think needs reworking. The formulation group sometimes say 'Look, based on our previous experience, and based on different molecules that the different groups work on, we think there is there is something better, so can you work on it more?' Alliance partners help to reframe as well. One director explains that working with a Japanese partner opened their eyes to compound types that they assumed were too complex to work with. As he put it, when they saw the new kind of molecule they said: 'Aw, I should have made that . . . It makes sense now but would I have done it? Probably not.'

Another scientist at a small biotechnology company points out that in the process: 'Very often you have very surprising findings . . . If you have a certain preparedness for new data and are willing to review and revise your hypothesis, you may find something that's completely new.' Another person said that when he hires scientists, he looks for good observational skills and the ability to 'look at what happens and maybe finding something completely different'. A final example shows reframing by drawing on unique expertise. A scientist tells about a group of specialists who look just at how to improve crossing the gut. He says they can come up with a novel conclusion about reworking the molecule in a particular way that they have heard about, but in this case they can influence the chemistry to make it soluble or increase the absorption in the body. The end result is that the patient does not have to be dosed with large quantities, and perhaps smaller quantities can show much better results.

Coordinating by Event-Time Pacing

The examples suggest a considerable ability for anticipation and appreciation across the different sets of experts, and considerable heedful interrelating as they pool ideas and intuition, iterate around their different perspectives on molecules and targets, and seek new ideas from their worlds of science that might help other disciplines. People try to stay actively engaged, since they cannot collectively follow clues or appreciate others' intuitive insights unless they are working concretely on the same thing. Understanding the need to reach out to others is one way to stay engaged, as this scientist explains:

> This is a drug formulation issue, so we actually have in house the expertise that can help solve those kinds of problems but you still need to reach out for that . . . it is going to internal meetings, sitting in on conference calls, social events, visiting people and talking about your problems . . . on the chance that they will say 'Oh you know what, I know this person that you need to talk to', and that happens all the time . . .

Many companies also try to encompass the entire cycle in the innovation process by going 'end to end' as this director explains:

> There are cross functional teams so we are trying to make it end to end as early as possible, so the team includes basic scientists, and early development, animal testing, and pre-clinical and early phase I trials and therapeutic experts. We bring them together to think about how things are going, how well development is going. It is a tradeoff because we do not want to squash creativity . . .

He raises the possibility of squashing creativity, perhaps by inhibiting some explorations of some alternatives, or some leaps to new trajectories through reframing. I suggest that by recognizing the need to cycle through the abductive learning routines, they can avoid squashing creativity.

However, the very long development times make it difficult for people to stay engaged end to end. The many kinds of specialists and numerous external contributors vastly expand the number and kind of relationships that projects need to coordinate. In one study, we suggest that project scientists can pace their work with emerging yet unpredictable learning events (Dougherty et al. 2013a). Event-time pacing takes advantage of emergence by using it to regulate the intensity and direction of so many people's efforts and attention. Emergence thus helps to coordinate as various groups stay attuned to the emergence of learning events in different development programmes. Learning events arise from the cycling of abductive learning routines, and indicate moments of closure in the development process as project scientists learn enough about their hypothesized configuration of interdependencies to indicate next thrusts in the development.

A few examples indicate that learning events do occur. First, a chemist working on the transition to manufacturing suggests a hypothetical learning event about a potential drug that seems to work well. She describes the next thrust into manufacturing and clinical trials, which opens up another set of challenges:

> Let's say we are successful here (with developing a potential drug) and now we are going to go to first in humans . . . In any development you have a lot of people working to make sure that happens. You have people from regulatory talking to our regulatory agencies, . . . from the pharmaceutics of the compound, the scaling of the compound, making the dry coloration of the compound . . . From clinical they are very critical and they are telling you how we are going to take this compound into humans and how we are going to do clinical trials; from toxicology and PK, pharmacokinetics, from legal, from commercial . . .

This event initiates a new stream of innovation activities.

A project team leader describes the learning event that moves a very early hypothesis into more complete development:

> We get very excited and sort of look at it from every different angle and get to the point where we say yes, this is something that is worthy of our efforts . . . sort of the final step of our process from original screening in the test tube is going to be, does this molecule behave the way you would like it to or think that it should in your best animal model.

This learning event involves considerable excitement, and arises because the molecule seems to behave according to their hypothesized configuration in animal models. They look at every different angle to evaluate their hypothesis.

Another project leader describes creating a good enough body of data around an early development learning event to engage the rest of the organization:

> Our job is to create a body of data and evidence around that hypothesis that engages the rest of the matrixed organization to get behind your idea and develop a drug.

A third team leader describes finalizing his drug development efforts so far and getting ready for early development by soliciting input from downstream units:

> We call that a contract meeting where we actually go to this early development committee and say here is the new drug, here is the kind of profile that I am going to make around it, here are the tests that I am going to do to show that it is safe in cardiovascular…when I have done all this I am going to recommend it forward so give me input now…….

He outlines his development issues and describes the other committees at the firm that are involved in overseeing drug discovery and judging whether or not learning events have occurred.

Finally, while only suggested so far, all firms invest considerable energy in overseeing the drug discovery projects to help figure out problems, anticipate hurdles, and consider whether or not particular projects still have potential. For example, this science director at a small biotechnology firm describes their oversight process:

> We make decisions at that committee, with scientists, mostly scientists, at the table also marketing participation, as to whether it is a good idea in the first place, and then we don't stop there. We then discuss, look at the project, what do we believe are the many hurdles in the initial stages, and what should the team that works on it address possibly first.

A committee of experts tries to anticipate hurdles and sort out priorities for project teams.

Managers of all the key scientific functions come together at a large firm as well monthly to review progress and problems:

> This is a monthly review of all (drug projects) at this Site. It is just managers and not the scientists themselves…we review the latest information and the idea is to guide the teams as to are they going in the right direction, are there ways we can help more…and then collectively the management team is looking at how to resolve those issues. Should the project stop, be accelerated, do we need more chemists do we need to get formulation a bit more involved? You have to be a little bit dictatorial at times, so if the writing is on the wall you have to make a management decision and at a certain point you have to say well, the odds of succeeding now are so low that we should move the resources onto something that has a higher probability of succeeding.

Project innovation in complex systems takes many hands and is very hard to do.

Persistent Problems with Taking Advantage of Emergence in the Project Subsystem

I have highlighted the complexities of innovation that the project subsystem in pharmaceuticals must deal with. The project-level complexity will not go away so innovators and managers must deal with it by taking advantage of emergence. We build the extensive use of intuition, luck, and clues that we found in our interviews into abductive learning routines that have already been briefly illustrated. I think that abductive learning coupled with event-time pacing help to organize, shape, and guide this long product development process. Scientists do not have clear answers, not even as projects go into clinical trials, because the unknowns and possible alternatives persist, so they keep navigating in the labyrinth of human biology with abductive learning routines. These inherent complexities lay the ground work for identifying other subsystems in the infrastructure for complex innovation systems that have other complex problems to address, also by using abductive learning routines. Scientific developments can enable navigating in the labyrinth, and strategic and institutional management subsystems create and shape the labyrinths in the first place.

However, my colleagues and I also find problems with using abductive learning and event-time pacing effectively. I summarize these problems here, and in the rest of the book explain how the other subsystems can help manage these problems.

The first set of problems comes from abductive learning itself. Academics and managers have not specifically developed this way of figuring out how to develop new products, and only our work attempts to detail the processes (Dunne and Dougherty 2016; Dougherty et al. 2013a; this book). Nobody we interviewed said they were using abductive learning or event-time pacing. The examples indicate how it would be easy to fall into competency traps, for example, since they rely heavily on experience, or skip one of the learning routines. We see fewer instances of reframing in our data (Dunne and Dougherty 2016), and speculate that people may be reluctant to cycle back to a different alternative. Several people also talked about heroic 'drug hunters' who could find drugs better than others, so intuition may be overemphasized.

Organizing challenges reinforce problems using abductive learning. People may lose sight of the collective learning objective to discover a configuration of interdependencies. Because there are so many specialties, each with its own infinite set of issues to follow and run down, sometimes project teams may treat other units not as co-developers but as answer machines. One bioinformatics expert told the story about being asked by a project team to find a biomarker for their cancer drug (biomarkers indicate binding to a particular kind of receptor). The bio-informatics scientist spent a great deal of time on the project and sent the project team 'lots of data', but he did not find a biomarker so the team 'was not happy'. He said that later on they realized that some of the receptors they found could be labelled, and that they were thinking too narrowly of a solution. The project team and this scientist did not reach out for clues and hypothesize how this finding might relate to their configuration of interdependencies, but instead tried to get an answer that did not exist. The bioinformatics scientist was not actively engaged in thinking through and evaluating the hypothesis, he was just asked to hand over answers.

Efforts to control the process may also weaken the collective intelligence by separating groups and trying to discipline the execution of steps. A biochemist contrasts being engaged in developing abductive hypotheses while working for a small biotechnology firm. That firm was bought by a large pharmaceutical along with several others, and folded into the very large, bureaucratic structure.

> [At biotech] I was very involved with the therapeutic area and cancer was my thing and I would propose targets and all that other stuff that a therapy team does and I read papers as well. So then we enter Pharma X and I am told well you are structural biology and you don't propose targets anymore . . . Another thing that does not work here is a free flow of information. Like down the hall, yadda yadda, we would eat lunch and something would come up and you would answer someone's question or they would answer your question, but that does not happen here.

These enormous organizations make it difficult to heedfully interrelate and stay engaged with concrete development efforts. External parties such as academics, small biotechnology partners, and other research specialty groups also contribute essential expertise to most development projects, but folding all these into the processes could be more difficult than connecting with internal units.

Second, resources do not necessarily flow smoothly in such a complex, long-term process. Some scientists noted that it could take managers many months to approve resources needed to explore or compare promising alternatives, so the scientists would just move on with other possibilities rather than wait. Third, project scientists also talked about efforts to exploit existing knowledge by following what the competition was doing, sticking to more familiar options, or focusing on problems they can solve. Some said that managers imposed 'tally marks' like the number of compounds synthesized rather than enabling innovation. Fourth, new technologies were continually introduced without careful implementation (Dougherty and Dunne 2012), so they disrupted development activities.

Tensions between project scientists, enabling technologists, and managers exacerbate these problems, I think, by inhibiting the abductive learning routines and event-time pacing. These tensions suggest that people are not attacking the knowledge system problems and strategic problems in the infrastructure. Instead, other scientists and managers seemed to be trying to solve their problems by imposing them on the projects. One director of a science group indirectly reflects the failure to address all the problems with this summary of the decision-making problems they face:

> [We need to] make realistic decisions at the right time. How much to invest in projects and when to stop them . . . The individual team member can be a passionate advocate for their compound and just show it in its best light . . . but at another level, to get better at making the decisions, [we need] better predictions of what is really going to make it and be something useful. Across a lot of the industry about 50 per cent of things that go into phase III trials [the most costly stage] never make it, and that is a shame. That is hundreds of millions of dollars and you should be able to predict better than that by the time you get to phase III. Part of it is figuring out how to use our data better and be better at predicting . . . and part is just making better decisions and not going with things that show marginal effects just for the sake of ploughing ahead and keeping the speed on . . .

He feels they do not assess projects very well or gauge their progress effectively. He also feels that they need better ways of managing the overall process so they can predict better—what I label knowledge subsystem challenges in the infrastructure. And he says that they make poor strategic choices and plough ahead regardless of prospects—what I label strategic management subsystem challenges in the infrastructure. Innovation projects cannot solve these problems, and in fact need solutions to those problems before they can proceed.

Tensions among Scientists and Managers that Inhibit Abductive Learning

I noticed from the beginning that different groups thought very differently about how projects should be managed and what they need to do to improve project innovation processes. Abductive reasoning generates relatively weak inferences about plausibility and provides no guarantees, while the need to generate revenues requires that new drug products be generated regularly. Almost everyone we interviewed is a scientist and familiar with the ordeals of drug discovery, but many people think that drug discovery projects worked ineffectively, and that ineffective project work alone causes the problems the industry faces with low R&D productivity.

To illustrate these contrasts, managers want to improve the drug discovery process. One R&D manager explains her challenges:

> I think that the hard part [for developing an ability to execute] is that it requires some discipline and process... I would submit that if scientists want to continue to flourish in the pharmaceutical world they are going to have to adopt some rigour and discipline that is part of a normal business environment, like keeping track of the time that they are spending on this project for instance.

We also see that some managers feel that the scientists are not making good decisions, as this R&D manager suggests:

> I wanted to... figure out a way to get the division [R&D] under control from a business standpoint, just literally be able to identify the parts and the pieces that would be levers that would allow us to get a handle on planning and improve our execution... We are putting pressure on the scientists to make better decisions...

I think the desire to 'get better' indicates that scientists and managers are not using abductive reasoning as well as they could. The project scientists' emphasis on clues and intuition no doubt worries the managers even more. However, as I detail in Chapter 5, managers' needs for less ambiguity and more new products prompt them to impose simple rationality on the projects, which most likely makes things worse on its own since simple rationality cannot work in complex innovation. My examples in this chapter indicate that some knowledge system scientists work well with project scientists. But they also clash with project scientists by expecting more methodical use of assays and other analyses. Some knowledge system scientists explained how their measures produce clear and useful information, while project scientists explained that much of this information is ambiguous.

The different groups have distinct issues and concerns, but rather than address these issues and concerns in their own right, knowledge system scientists and strategic managers push them onto the projects. The projects cannot resolve these non-project challenges. I analysed these differences to highlight key tensions that may pull innovation

Table 1. Contrasting Foci and Approaches to Drug Discovery

Innovations problems	Project scientists	Knowledge system scientists	Strategic managers
Project focus and approach	• Realities of compound and target • Searching with clues, intuition for configurations of interdependencies	• Methodical measures of steps • Delving deeply into steps, generating data	• Indicators of progress, efficiency • Structuring decisions
Process management focus and approach	• Applying science to particular issues • Unique, varying behaviours • Finding paths	• Optimizing process • Invariant aspects • Finding answers	• Reducing uncertainties sooner • What we know now • Keep ploughing ahead
Strategic management focus and approach	• Building confidence, reframing • What we do not know • Arriving, approaching	• Testing and verifying • What we can know better • Measuring, watching	• Balancing pipeline • Aligning resources w/best opportunities • Finding cash flows

efforts apart, and inhibit abductive reasoning. These clashing concerns show that people in this infrastructure need to acknowledge that there are qualitatively different problems that need to be addressed on their own terms, not by the projects. Table 1 contrasts the perspectives starkly to emphasize points of conflict (although in fact more overlaps exist).

The first row on project focus and approach shows qualitatively different ideas and expectations. Project scientists focus on the concrete realities of particular compounds and targets and use clues and intuition to imagine hypotheses that might enable them to learn more about the interdependencies between them and the body. Knowledge system scientists in specialty groups think more about careful measures of the steps, while strategic managers worry about progress and efficient development. Knowledge system scientists may be right that some project scientists do not use the latest assays and other techniques. My main point is that the project scientists do not measure steps precisely or generate clear indicators of progress because they cannot. But the other two groups expect these outcomes and so are alarmed by what they see going on in the project efforts. The innovators together in the infrastructure can develop measures and indicators of progress, but these will be new and will fit abductive learning.

The second row contrasts the different perspectives on the innovation process. Project scientists apply their science to particular issues in the discovery process and focus on the unique and varying behaviours of their hypothesized configurations to find paths forward. Knowledge system scientists want to optimize the process in general so they focus on invariant aspects to find answers about best procedures. Strategic managers look at the process as reducing uncertainties, and focus on what they know now to push forward. Again, knowledge system scientists and strategic managers expect the project scientists to address different activities than they do now address, or can address. The innovators cannot optimize each development project based on generic standards, but they can develop better strategic paths. They may not be able to reduce uncertainties sooner, but they can enlarge the infrastructure's capacity to use more alternatives.

The third row contrasts perspectives on strategic management. Project scientists move stepwise and build their confidence as they proceed, and occasionally reframe their ideas. They emphasize what they do not know and are always arriving and approaching, not hitting the final objective. Knowledge system scientists want to test and verify and concentrate on what they can know. Strategic managers want to balance the pipeline and align resources with the best opportunities, and find cash flows. The innovators need the entire infrastructure to verify, and to align resources.

If we look diagonally in the table from the top left to the bottom right, we see three very different problems and solutions that do fit different subsystems. Project scientists need to concentrate on the concrete realities of possible drugs, since they cannot abstractly analyse or find anything without looking carefully. Knowledge system scientists are right to try and optimize the process over all, but they would do so by integrating sciences into strategic paths for projects. And strategic managers are right to want to balance the pipeline and align resources, but they would do so by developing a richer portfolio of value-creating opportunities that reaches far into the future.

Conclusion

Participants in the project subsystem of the infrastructure of complex innovation systems play a central role. Project innovators actually build the products, services, or programmes that solve concrete problems that are part of the broad infrastructure objective. I emphasize the actual building and actual solving of real problems, because doing so is difficult even for incremental new products. People who do not actually build products may fail to recognize the detailed and elaborate challenges of constructing a functional product. To build a complex product, innovators need to first figure out what might be involved in the building and how, and how might the various components go together to produce the whole. They navigate in the labyrinth of the complexity, and for new drugs, in the 'living labyrinth' of human biology (Barry 2005). But people in the pharmaceutical infrastructure who do not actually have to navigate in the living labyrinth of human biology may not recognize the inherent complexity of this innovation process. I have suggested that some knowledge system scientists and strategic managers want the drug product innovators to stop navigating and proceed directly to the destination.

We suggest that abductive learning provides the basic process for figuring out how to create complex new products (Dunne and Dougherty 2016), and that event-time pacing provides the process for coordinating so many people over such a long period of time (Dougherty et al. 2013a). Abductive learning and event-time pacing constitute two of the new social technologies that enable innovators to make use of the enormous potential in the explosion of new biomedical sciences, despite their nascent and very emergent nature. Both social technologies transform project innovation processes so that they can take advantage of emergence. I developed abductive learning routines in Chapter 2, and illustrate how innovators use them to take advantage of emergence in this chapter. Our papers in the References provide more details.

One important message in this chapter is that project-level innovation cannot proceed well unless the other subsystems in the infrastructure simultaneously address their own distinct complex problems of innovation. Without better strategic paths for drug discovery, project innovators may stick with incremental exploitation, especially if managers demand faster development. Without enhanced strategic options for alternate applications, project innovators may discard potentially useful products because they might require new delivery mechanisms or associations with healthcare systems for their use.

This chapter indicates that the failure to recognize and deal with the inherent complexity also inhibits effective project innovation in complex innovation systems. Because they are unfamiliar with abductive reasoning, people in the complex innovation system may impose conventional approaches, which is why I think that R&D productivity is declining. Conventional approaches to reasoning and simple rationality cannot work, and most likely make things worse than they are.

Unfortunately, abductive reasoning is messy and unpredictable during the process. Clues help codify the noisy information in complex systems, while intuition enables people to imagine configurations among interdependencies that might become a drug. Focusing on clues is hard because clues highlight what they do not know. But when clues are not used to direct the navigating, innovators and/or the managers may ignore a good deal of the information that is available. Evaluating by elaborating and narrowing keeps innovators open to possibilities while stabilizing the process, but a drive for clarity in the early stages of projects may push scientists to stay with current results rather than reach out to learn more. Reframing by iteratively integrating across disciplines and experimental modalities pulls together the perspectives and conflicting ideas into a reframed hypothesis that makes sense to most. But efforts to weed out bad ideas as early as possible can weaken the reflexive deliberation that combines ideas.

4

The Knowledge Subsystem in the Infrastructure for Complex Innovation Systems

Designing the Strategic Path for Drug Discovery by Integrating Sciences and Technologies

The second subsystem in the infrastructure of complex innovation systems encompasses the numerous fields of science and technology that can enable and inform innovation projects. For drug discovery, an astonishing array of scientific and technological disciplines contribute to this innovation. A short list includes advances in genomics, microbiology, biochemistry, pharmacology, medicinal chemistry, physical chemistry, and biology; bioinformatics, structural and functional genomics, combinatorial chemistry, systems biology, proteomics, and metabolomics; recombinant DNA, SNPs, genetic engineering, monoclonal antibodies; and new imaging approaches (NMRs), and high throughput screening. According to Pisano (2006), all these new sciences and technologies have vastly expanded the range and type of agents that can be used as drugs, the number and type of drug targets, and methodologies to design and build drug products. However, to take advantage of emergence, all these new sciences need to interrelate and link up so that deviation amplifying feedback loops can generate different approaches that may work better, and so that new possibilities can be combined, reused, and recombined over time.

We saw at the end of Chapter 3 that the project subsystem cannot work on its own to figure out how a potential drug will behave in the body against a disease, because so much is unknown. Project scientists are navigating in a multi-dimensional labyrinth as they discover and develop drug possibilities. Their work can easily unravel from searching for a configuration of interdependencies to focusing on each part separately. Project scientists need help with imagining viable configurations in the first place, with evaluating alternative moves as they proceed, and with reframing their hypothesized configurations to accumulate learning. When asked what he thought would enhance the drug discovery process, one chemistry team leader said:

> I think the biggest problem a discovery person and an early development person faces is how can you design your strategic path so that you can make those decisions faster and not spend resources on drugs that are not going anywhere.

Building on this comment, I define the core problem of this knowledge subsystem as designing and redesigning the strategic path for drug discovery and development in diverse diseases. Designing the strategic path involves finding good starting places through systematic and integrated early screening, choosing among alternate paths in the labyrinth, and transforming underlying disease paradigms so that the discovery process is enhanced. Designing the strategic path requires the ongoing integration of sciences and technologies around the three abductive learning routines and heedful inter-relating.

Scholars and managers created new social technologies to build knowledge systems to support the many new projects in the 1980s and 1990s. Scholars and practitioners alike recognized the need to not only manage innovation projects as effectively as possible, but also to build the long-term capabilities in product technologies, science, marketing, and other knowledge systems so that particular projects can more effectively get to market. A spate of books and articles heralded this transformation in innovation management by developing new social technologies for long-term knowledge development. For example, third-generation R&D emphasized the transformation of the R&D function from a separate activity to a long-term strategic connection with businesses (Roussel et al. 1991). The marketing discipline emphasized understanding new customers and markets (Day 1990), and operations developed complex social interrelations among the scores of new projects ongoing in any major organization (Cusumano and Nobeoka 1998). The gist of all these books is that new social technologies—new ways of thinking and acting—are needed to enable continuous new product innovation by building long-term capabilities in technology, sciences, marketing, operations, and other functions.

Transforming Knowledge Subsystem Social Technologies for Complex Innovation Systems

Another transformation in social technologies is necessary to build the knowledge systems for the infrastructure of complex innovation systems. This transformation changes how we understand the role of basic and applied sciences and technologies so that they can continually integrate their ideas into strategic paths for innovation. Industry experts have already defined what needs to be done. First, many show that knowledge is literally in the network, not in any one agent or agency (Owen-Smith and Powell 2004). In addition to networking, Collins (2011: 1) argues that translational science needs to be re-engineered 'to catalyze the generation of innovative methods and technologies that will enhance the development, testing, and implementation' of new drug therapies, devices, and diagnostics. To do so, he says it is necessary to shift from one-off solutions towards a more comprehensive strategy that would combine sciences to enhance broadly applicable abilities such as target validation or expanded chemical space. Pisano (2006) points out that each new science stretches the search landscape, and each is like shining a light on one aspect of the problem. To discover and successfully develop a drug, it is necessary to illuminate the whole problem through the joint optimization of all the sciences in an integrated fashion.

We know what has to be done, but how to do so is less clear. I propose that cycling through three abductive learning routines generates long-term knowledge development and integration into strategic paths for ongoing taking advantage of emergence. This chapter explains how and why that can happen. Two additional changes are needed to transform the role of science and technology for innovation to accommodate complexity.

One transformation is to move away from the nineteenth-century belief in prediction. Nightingale (2004) synthesizes many ideas on the evolution of science and technology into a framework based on the idea that unpredictability is pervasive. The world is not predictable, and science does not discover the natural predictability of the world. The notion of prediction comes from nineteenth-century beliefs that there is a universal, deterministic, and machine-like causality that is to be discovered by science. Instead, scientists construct predictability as they generate better explanations through repeatedly intervening in the world and modifying the categories they use to understand it. To construct predictability, scientists iterate with technologies to create artificially simplified experimental conditions where theory and reality match.

Scientists use these experimental conditions to contrast the divergent implications of competing explanations. This ongoing iteration between and among sciences and technologies requires what Nightingale terms a 'largely overlooked' infrastructure of skills, socially distributed explanations, and physical objects that are common to both.

Integration of sciences and technologies designs and redesigns the strategic path for ongoing innovation in the infrastructure for complex innovation systems. In the case of pharmaceuticals, integration is essential to shape and guide the complex 'grunt work' of project-level innovation to discover and develop new drugs. We suggest that project scientists navigate in the labyrinth of human biology to develop drug possibilities (Dunne and Dougherty 2016). Knowledge system scientists need to construct the means for navigating. However, integration is not now adequate because the paradigm of science as prediction still prevails, so people seek to eliminate uncertainty, which ignores the need for social technologies to make new sciences and technologies useful.

Second, it is necessary to rethink the role of the ideas of science and of scientists in integrating ideas. Rethinking does not require a drastic change since, while we tend to think of basic science as detached from applications, Nelson (2005) and Stokes (1997) point out that much of science is already applied, in that scientists want to solve real problems. However, existing systems of sharing knowledge emphasize the creation of novel understandings. These novel ideas are presented to others by publishing, peer reviewing, grant proposing, attending conferences, making presentations, patenting, and so on. The change is to realize that new ideas are clues to possible strategic paths, not solutions in their own right. Scientists need to think about interdependencies among ideas. Every scientist need not figure out how her new idea interrelates with all the other ideas, but each can think about how her new idea could work together with some others to illuminate a strategic path for innovation. Abductive learning routines centre on 'T-shaped' skills for interrelating, whereby people are 'not only experts in a specific technical areas, but also intimately acquainted with the potential systemic impact of their particular tasks' (Iansiti 1993: 139). Leonard Barton (1995) argues that people possessing these T-shaped skills are able to shape their knowledge to fit the problem at hand rather than insist that the problem appear in a particular form.

Examples will help illustrate how science provides clues for imagining a configuration of interdependencies among ideas that can design a strategic path. In *Technology Review*, Hall (2014) and Rotman (2014) describe new tools and sciences that can revive the moribund effort to improve treatment for psychiatric disorders (e.g. schizophrenia, depression, PTSD). Little or no progress has been made since the 1960s with these disorders, because trying to understand the enormously complex circuitry of the brain is, well, enormously complex! One new technology is optogenetics (Hall 2014), which allows scientists to turn on or off specific neurons in the brain and explore if these particular connections are responsible for a behaviour or disorder. The scientists genetically modify neurons so that they produce light-reactive proteins. Using a hair-thin fibre optic thread that is inserted into a living brain, they use blue light to activate particular neurons, and red light to silence other neurons. This technology combines optics with genetics and genomic tools to create a new more detailed way to understand disease. According to Hall, this emerging strategic path can reinvent the way brain science is done.

Another new technology to learn more about brain disorders is to take skin cells from people with, for example, schizophrenia or autism, turn those cells into stem cells, and then nudge the stem cells into functioning brain cells (Rotman 2014). Scientists can directly examine in molecular detail what is going wrong in the brain cells of patients with these conditions. When these brain cells are integrated with genetics created by fast and cheap DNA sequencing and high throughput screening of molecular compounds, scientists can explore possible approaches for drug therapies. Rotman quotes the director of NIMH who said that the old way was to tweak existing drugs that were discovered by accident in the 1960s. The director said that this approach was 'a crude strategy' that ignored the underlying biological mechanisms. 'By studying the drugs, we were led down a path that said depression was about being a quart low in serotonin . . . But that isn't how the brain works. The brain is not a bowl of soup; it is really a complex network of networks.' These new understandings are opening up new strategic paths for brain disorders. Rotman (2014: 40) notes that 'designing drugs to precisely target circuits in the brain' is a distant opportunity, but it is what I would term a new strategic path.

Third, '(f)aulty genes have a significant role in causing brain disorders' (Rotman 2014: 38). The conventional approach to drug discovery is to identify the gene related to that illness, and test compounds against the protein it codes for. However, most psychiatric illnesses are caused by combinations of genetic variants. Rotman reports that Pamela Sklar, director of the department of psychiatric genomics at the Mount Sinai School of Medicine, thinks that the numerous variants provide more chances to home in on key pathways. The different variants may activate common sets of pathways so they together can help pinpoint what is going wrong in particular diseases by taking the interdependencies of the disease into account. Sklar also noted that large gaps in understanding remain, and 'with so few pieces of the puzzle, it's still hard to know how it all hangs together'. But she and others are trying to figure out 'how it all hangs together' by integrating different sets of insights.

These examples illustrate Nightingale's (2004) contention that science and technologies iterate with each other, and I would emphasize integrate as they do so. Science is used to create technologies (adapted stem cells) that are used to create science (new insights into brain circuitry and disease). The physical artefacts (stem cells that create diseased neurons) create the conditions where competing explanations can be selected, and resulting explanations can create new physical artefacts.

There are many other examples of integrating sciences and technologies. In our study of the use of new sciences and technologies within pharmaceutical companies (Dougherty and Dunne 2012), we show that people are piecing together different insights from, for example, combinatorial chemistry, NMR, databases from toxicology, and HTS to help project teams. Dunne (2015) examines the array of alliances in pharmaceuticals, and emphasizes the critical new knowledge arising from what she calls truly innovative alliances, where different parties interweave distinct sets of expertise. Su (2013) examines ways to enhance the links between basic and clinically applied research, and details a variety of new kinds of collaborations that are forming to overcome the persistent challenges in the academic-industry connection.

Barriers to Integrating Sciences for Complex Innovation

However, all these studies also show that actually integrating the emerging sciences and technologies with drug discovery and development remains very difficult. Our research suggests that people in the infrastructure do not use all the new sciences and technologies effectively because they are still hung up on the notion of prediction, and on the idea that science produces solutions that fit right in, so we do not need new social technologies. Several examples from our interviews with discovery scientists show that they already had an understanding of the strategic path: it is based on an assembly line of distinct steps to be executed one by one in sequence (Ng 2004). They also already had an understanding of the role of new sciences and technologies: to scale up and speed up existing steps. Unfortunately, these erroneous understandings were and perhaps still are being reinforced by industry experts and academics, including management scholars.

A research director reviewed the history of the biotechnology revolution as follows:

> In the 1980s it was rational drug design, and in the 1990s it was combinatory chemistry and genomics. They said these things would fundamentally change drug discovery, they thought these would be the architecture. So in the late 1990s they thought that by knowing the sequence of all genes it would be better, we would get faster. But sequencing all the genes has not helped. Most genomics companies have gone belly up.

He points out the repeated waves of breakthrough sciences over several decades that were supposed to fundamentally change drug discovery, but did not. The underlying belief was that the new sciences were themselves better and faster than traditional pharmaceutical R&D.

A vice president of enabling technologies recounts a similar but more detailed history around combinatorial chemistry:

> What happened originally is we tried to do everything by huge numbers. In the high frequent screening . . . age in the early '90's everybody thought more was better. What happened is that little companies sprung up and they would publish their first papers in the *Wall Street Journal*, which would get the CEOs all upset and say our chemists or biologists are not up to it so we have to get more productive . . . With high frequent screening the molecular biology revolution allowed us to be able to clone and express things pretty quickly. What happened is that we try to do everything on an extremely large scale. And what we found was the compounds that we were screening and the compounds that we were making with combinatorial chemistry, we really did not have a clear view of what they are . . .

Everyone thought that 'more was better'. By increasing the amount of chemical compounds that were screened against targets they would find drugs more quickly. This linear view says that the existing approach was fine, it just needed to be amplified. Unfortunately, complexity got in the way. He also points out that companies felt they needed to catch up with biotechnology. He goes on to note that the molecular biology revolution did not result in much progress, since trying 'to do everything on an extremely large scale' did not address the inherent complexity. They mistakenly assumed that scaling up existing steps and plugging in the magic of new sciences and technologies would save the day, but it did not.

Another VP of technologies at another big pharmaceutical said the same thing about combinatorial chemistry, explaining that they did not really understand the chemistry:

> Everyone thought you could make a million compounds from tossing lots of different pieces together and hoping for the best to screen them. You find it is

not that easy and you have to do lots of traditional chemistry making molecules that are not amenable to that type of chemistry.

As he says, they learned that they needed to integrate new technologies with traditional chemistry. It seems that everyone black-boxed the actual and onerous processes of drug discovery that I detailed in Chapter 3.

A chemistry vice president at a third pharmaceutical was more blunt about their initial but mistaken expectations:

> One of the changes over the last 15 years or more... what happened is that we went to a time when we thought that everything would be high through-put screening. We thought that by sheer numbers a gem would pop out. No, that didn't happen. Now we are smarter. It is an iterative process. First you get the data and then redesign the chemistry and rethink it... We must observe, so we cannot bank on an initial set of data. There is no crystal ball. And you can't run all through a half a million compounds in all tests. Early tests are 10,000. And no assay is 100 per cent predictable, so we take our best guesses and push a fraction forward.

He highlights their return to iterative, more focused, hands-on drug development and away from the 'crystal ball'.

A bioinformatics director assesses the initial uses of genomics in the same way:

> There are extreme models of how an organization responds to a new technology, for example, genomics, very little came out of this... How could management make decisions about these technologies? This investment had no clear proof, the irony is that it could become more practical now that the investment has evaporated. It was hype... there was no account-ability, they didn't think through downstream metrics, it was the number of data points generated and they didn't make anything... Genomics has a specific role, but it was a gross over estimation... The initial surge was that we won't need to do clinical studies... They thought they would find new drugs (and) they thought they had chips that would do the clinical tests, but it turns out to be more complicated than that.

'(B)ut it turns out to be more complicated than that' summarizes their initially erroneous attempts to apply all the new knowledge. The initial belief was that the new sciences and technologies would simply replace existing pharmaceutical R&D practices. Rather than embrace the inherent complexity and figure out how to use all the new ideas to take advantage of emergence, the biopharmaceutical industry tried to scale up specific steps—more is better, speed is better, and new sciences provided solutions. However, as this bioinformatics director notes, they are figuring out how to use genomics and what its particular role can be. Unfortunately, investment had dried up.

In our study (Dougherty and Dunne 2012), we show several examples of how scientists inside the firms are learning to integrate the new sciences and technologies and combine them with their existing science. For example, the second person quoted regarding originally trying to do everything by huge numbers explained that now they evolved by integrating new sciences like combinatorial chemistry with mathematical tools and traditional approaches.

> So then what happened is there was an evolution that instead of super huge numbers of compounds, like screening 100,000 compounds a week we would try to cull down our library and use some mathematical tools to cluster what our structures look like in the library. . . . So that actually gave us more quality results because basically what you are doing when you are taking a high frequent screen is you are drawing a line and saying if this has this level of activity then we will look at it but if it does not then we will not look at it. Now we have some mathematical tools that allow us to cluster around our activity to be able to do more with it.

They recognize that these solutions did not work as initially presumed, and are integrating various sciences and technologies to address the integral problems they face. Integrating the sciences was still a work in progress when we finished the interviews. More recent updates indicate that firms, academic institutions, and other agencies are learning more about integrating the sciences and technologies (e.g. Collins 2011). But the innovation system is still trying to fold in everything effectively, and the industry has significantly reduced their investment in R&D.

Industry scientists and managers were not alone in the belief that the new sciences and especially biotechnologies would save the day by providing solutions. The promise of biotechnology was also hyped extensively by many scholars and commentators, including those in management, and in science, technology, and innovation. The new sciences were said to be more precise and more 'scientific' in that they would identify actual mechanisms of drugs. According to Pisano (2006), these new sciences were assumed to transform the economics of pharmaceutical R&D by enabling a less random, less uncertain drug discovery process. Drug discovery was not complex, it simply lacked information, and biotechnology would fill that lack. There was no need to learn to take advantage of emergence by designing or redesigning strategic paths, since biotechnology would remove the complexity. Instead of using knowledge to take advantage of emergence by finding paths through the 'living labyrinth' of human biology, they tried to fill the bottomless pit of the infinite information gap.

The biotechnology revolution was a myth, according to Hopkins et al. (2007: 566). This 'myth' was that 'biotechnology is transforming pharmaceutical innovation by increasing the number and the effectiveness of drugs and diagnostics' (Hopkins et al. 2007). To illustrate the many presumptions behind the 'revolution', Arora and Gambardella (1994: 526) say:

> Rational design of molecules is gradually replacing random, trial and error experiments . . . Growth of scientific understanding in molecular biology and genetic engineering has clarified important aspects of human metabolism and the chemical and biological action of drugs . . . by studying the structure of receptors, scientists can design (typically on computer) a theoretical compound that matches a given receptor site and is expected to counter a certain pathology.

Our quotations indicate that 'it's more complicated than that'.

Another illustration comes from Henderson et al. (1999: 267), who say: 'The last 25 years have seen a revolution in the biological sciences that have had several dramatic effects on the global pharmaceutical industry.' According to Powell et al. (1996), biotechnology is a competence-destroying revolution because it builds on a scientific basis that differs significantly from the knowledge base of the more established pharmaceutical industry. These authors highlight rational drug design and molecular biology that scale up the production of protein targets for the high throughput screening of hundreds of thousands of molecules against each target. According to Scannell et al. (2012), however, drugs may derive their therapeutic benefits from interactions with multiple proteins rather than a single target. High affinity binding to a single target—what rational drug design produces—is limited because the causal link between targets and disease states is weaker than commonly thought. We see considerable 'hype' about the wonders of biotechnology in many other management publications.

The 'hype' went far beyond management research (Dougherty 2007). Many can recall the human genome project, the results of which were announced in June 2000. The expectation was that sequencing human DNA would reveal the genetic causes of disease and lead to diagnoses, treatments, and cures for intractable illnesses like many forms of cancer (Cohen 2011). It has since become clearer that a variety of other factors, once thought minor, are as important to our health as genes themselves. A person's susceptibility to disease may depend more on the combined effect of all genes in the background than on the disease genes in the foreground. The growing list of common diseases that have been traced to multiple genetic variants includes everything from types of blindness to autoimmune diseases and metabolic disorders like diabetes. Brain researchers have also identified hundreds of genes associated with schizophrenia, and some scientists think the number could go as high as a thousand (Rotman 2014). New studies reveal that the entire framework of medical taxonomy requires rethinking and that therapeutics of the future will likely be designed with cellular networks in mind, rather than being limited by historical designations of disease category (Collins 2011).

The continued decline in R&D productivity despite all these breakthroughs belies these various promises (Collins 2011; Scannell et al. 2012). The number of compounds that made it to clinical testing has not markedly increased despite the industrial R&D from the scaled-up steps (Pisano 2006: 96). Many biotechnology companies have either failed or shifted business models from selling technology platforms like genomics and proteomics to becoming vertically integrated themselves (Sammut 2005). According to Scannell et al. (2012), more of the first in class molecules approved between 1999 and 2008 were discovered using phenotypic assays than using the target-based assays based on molecular biology. Since target-based screening dominated, one would have expected more target-based discoveries. These authors suggest that microbiology-based screens may be efficient for pursuing validated therapeutic hypotheses, but they do not facilitate the search for innovative drugs.

All these authors are right to emphasize the enormous potential of all these new sciences. But, as Hopkins et al. (2007), Pisano (2006), and Gittelman (2015) argue from the available data, there has been no revolution, only an evolution. The new sciences and technologies do not replace traditional medicinal chemistry, pharmacology, and physiology but rather must work with them. The biotechnology 'revolution' did not arrive because 'it's more complicated than that', to requote the director of bioinformatics. The complexity of human diseases and biology did not, and will not, go away. Pisano (2006: 36) points out that trying to discover drugs by identifying all the genes and proteins in the human body with the scaled-up processes is like trying to fly a plane by listing all the parts.

According to Hopkins et al. (2007), inappropriate models from academia have led to inappropriate investments and policy decisions. By abandoning the revolutionary model as a myth and promoting more realistic expectations around the incremental, complex nature of major technological change, they argue that it may be possible for policy makers to promote the development and adoption of biotechnology. Nelson (2005) also points out that the actual paths to application of apparently promising scientific discoveries are in fact uncertain. Learning to use new technologies takes considerable ingenuity, because unforeseen capabilities and uses are discovered en route and different technologies interact in complex and surprising ways. The way that firms organize their implementation of a common, broad technology such as internal combustion engines makes a huge difference.

I argue that the best way to promote the development and adoption of new sciences and technologies into systems of complex innovation is to integrate them with one another and with the actual innovation process, and use them to take advantage of emergence. Rather than assume that the new sciences provide solutions, assume that they provide better clues, and better alternatives to explore. Rather than try to fill in the infinite information gap, use available knowledge to search more productively. Rather than try to generate millions of facts about molecules, biologics, genes, proteins, and other biological processes, try to figure out which facts are relevant in particular situations.

Abductive Learning to Integrate Science and Technologies for Innovation

All these new ideas do not provide answers or eliminate complexity, but they can significantly enhance innovators' abilities to take advantage of emergence if these ideas shape and frame the innovation process by designing the strategic path. Participants can use the sciences not as a source of answers or of solutions, but as part of the construction of experiments that create the conditions for predictability. According to Nightingale (2004), scientists contemplate results, reframe their problems, and enter iterative problem-solving cycles with technologies that involve tinkering with the technologies and modifying categories until ideas and the world can be made to match. New sciences and technologies provide ways to understand and to frame the problems, ways to intervene in the world to uncover what might be going on, and ways to compare and contrast divergent implications of competing explanations.

I propose that people in the knowledge subsystem use the same learning routines as applied in the drug discovery projects, but now use them to address the challenges of integrating the sciences and technologies to continually design and redesign the strategic paths for various drug discovery areas. The three abductive learning routines enable participants to do so by cycling repeatedly through formulating, evaluating, and reframing hypotheses about the configuration of interdependencies among sets of sciences and technologies that can help create useful strategic paths.

Reframing the Configurations of Interdependencies among Sciences and Technologies

I begin the discussion of the abductive learning routines for drug discovery with reframing. I have marshalled arguments that sciences and technologies are not stand-alone solutions, and that new social technologies of abductive reasoning are needed to integrate them effectively. Both these ideas require significant reframing of practices. While, as suggested, the reframing is under way, these big transformations are difficult—and they cannot occur unless all four subsystems are transformed.

But people recognize that something is seriously wrong. Many of the scientists and managers we interviewed said that their approach to innovation was wrong, but they were not sure about how exactly to transform the process. For example, a scientist quoted in Dougherty and Dunne (2012) said:

Making a decision about what to work on is the most critical.... A consultant... said 90% of people are working on the wrong thing, failure is so great. We don't know what is right. It will not change, it will get worse. The attrition will be greater but at earlier stages. The attrition will be more invisible. Need to make the right decisions so if you can stop something early... we are weeding out early.

While he does not say they need to redesign the strategic path, he does suggest that the whole industry recognizes that their current approach is not working. I suggest they need to take the next step and reframe their approach to find better ways for using all that they do know. A director of biology told us that with all the technology tools they can find lots of potential molecules. But now they need ways to make better decisions about moving molecules forward.

A scientist at another large pharmaceutical firm delved more deeply into what he thought was wrong with their process:

I actually think our model is broken right now. It is across the industry. The real challenge right now is that we are so process heavy, so layered in the way that we develop molecules that we have systems that we expect our molecules or projects to conform to. But this being biology frequently you cannot just fit the program into a fixed and flexible system...(do you mean clinical trials?) Yes and our decision making processes and our governance and the way we make decisions about how the trials can be designed, what the end points are ... The problem is that the organizations that have been most successful have actually kind of avoided some of these highly structured predefined models where we try to learn from our past experiences and apply those rigorously to all future efforts ... We try to get a one size fits all solution for a portfolio of biological projects and frequently it is not a good fit ...

In this scientist's opinion, their process now is too rigid, and they impose a linear and prediction-based model that they 'expect our molecules to conform to'. Indirectly he acknowledges the inherent emergence where a one size fits all model cannot work, and he also said that the projects 'demand unconventional development pathways'. His new model would be to allow separate development units around different diseases to work independently. He said they should be fully resourced with the right people, funding, and be big enough to have the critical mass to carry out this complex process, but freer to customize their processes.

Clearly the industry needs to break out of its existing procedure overall. My emphasis is to put new and old sciences and technologies together in new ways. The knowledge subsystem needs a paradigm shift, to stop looking at elements such as targets and molecules apart from how they function to generate the disease, and start thinking of all the new sciences and technologies as clues that can lead them out of perplexity. Small steps can begin the reframing, such as appreciating that single breakthroughs, no matter how cool, have little value alone. The real question is what does this breakthrough depend on to help define choices and make better decisions? The knowledge subsystem overall has to surface and challenge assumptions that lead to the wrong process, and explore how they might rethink existing interdependencies in the face of new, emerging knowledge. The new directions in brain science mentioned before indicate how they switched from focusing on increasing serotonin, for example, to exploring networks of neural connections to understand the disease process. Scannell et al. (2012) suggest another simple reframing—move away from the high throughput screening (HTS) alone since it does not work and recognize the usefulness of directed iteration by chemists. They propose that companies learn to combine the advantages of HTS with the advantages of small teams.

The technologies and sciences together can help map out possibilities and good alternatives, not simply industrialize steps, so over time people try out different configurations that may lead to new performance objectives that reflect new alternatives and consequences. Reframing is modifying the categories, and we see in the examples that people need to modify the expectations for prediction and for eliminating uncertainty. They need to reframe decision-making by asking: are we doing a better job of shaping search spaces and finding good alternatives? Are we now learning enough about particular projects to take the next step in development?

Formulating Hypotheses: Using Clues to Imagine Configurations of Interdependencies among Sciences and Technologies that Design the Strategic Path

Reframing the approach overall to drug discovery and development opens the process up to new configurations. With some opening up to new possibilities in the knowledge subsystem, participants can begin to use new scientific and technological insights as clues to imagine new configurations of interdependencies that design the strategic path for particular diseases or networks in human biology. Thinking of the new ideas as clues rather than as solutions helps people to use the ideas to lead them out of perplexity—to imagine a world in the discovery process in which that clue is meaningful (Weick 2005). Knowledge system scientists are constructing predictability with their imagined configurations. They are imagining the artificially simplified experimental conditions that will allow them to explore the divergent implications of competing explanations for ameliorating a disease with a drug. To construct predictability, it is essential to understand what any new idea depends on in order to be useful. The goals are to generate alternative possible paths, and look further into those alternative paths so that project scientists can make better decisions for going forward.

Several examples illustrate the idea of integrating sciences and technologies to design the strategic path, and show that people in this complex innovation system are attempting to do so. First, one director of enabling technologies explains how they have reorganized the process of choosing specific drug projects with an 'open proposal process' that includes input from everyone at the research site:

> Disease area experts present a proposal to all the employees at the site. It used to be senior leaders. The proposal is what target, how it connects to a disease, and what do they need to move forward (like) develop the protein, do lead generation chemistry, structural biology. We have a small body of technical team leaders and biology and chemistry team leaders. We discuss, review and make a decision. If we say we will not support the project because there may be some technology gap or something is not well developed, so you need to do x, y, or z, and they may need the technology group to help with that too.

> It can go back to the scientists and they can ask for help for the technology group. If yes, we will work on this particular target, and know what we need to do, if we have to make a protein how will we do that . . . By going through the open proposal process, we can identify brick walls and ways to get around them.

This example is just a beginning, and focuses in on the project itself rather than on the strategic path for development. It does not involve the abductive process of imagining a configuration of how these various technologies and sciences construct a possible strategic path. But it shows that they are putting the puzzle pieces together as heedfully as they can.

Another example concerns bringing assays forward in time rather than proceeding sequentially one by one. We found the idea of bringing assays forward throughout our interviews. I interpret this as another indicator that integrating the sciences and technologies holistically is occurring already. A VP of technologies explains:

> We are doing a lot more of the safety testing too earlier on. All the parameters around [the] physical/chemical properties of a drug and the pharmacological properties of a drug are being gathered earlier. In addition to that we are using genome wide approaches to validate targets with our knockout mice to query drug activity. We are applying our drugs to cells and we are applying our drugs to animals in toxicology or efficacy studies and we are examining tissues from those animals or fluid from those animals in systems wide ways. We are using gene chips to look at how all the genes get modulated in the liver of animals in some ways or . . . we are looking at the urine in an NMR machine and looking at the metabolites in that urine as a global profile and how that changes with treatments, and that is called metabolomics . . . My group . . . is collecting lots of data, handling it in some way and analysing it with statistical tools and visualization tools to help communicate (results to the project scientists).

In this example we see a variety of technologies working in concert to generate tests and integrate diverse sets of information.

More examples come from Dunne's (2015) study of alliances between different firms in pharmaceuticals. She finds that new ways to continually redefine the project goals and to continually recon-struct the capability to interweave expertise distinguish truly innova-tive alliances. The truly innovative alliances integrate their different knowledge into a new pattern that allows the participants to see a new path for a particular disease area. However, most alliances in her study were transactional. Firms simply handed off a task to a partner to complete, and did not learn from the partner about how they carried out the work.

Su (2013) studied a variety of new kinds of collaborations between academic researchers in universities and clinically applied researchers in pharmaceuticals or not-for-profit research associations. One is a large pharmaceutical firm's Center for Therapeutic Innovation, which solicits proposals from academics for drug possibilities that need pre-clinical and clinical tests. The company selects projects that fits their portfolio, which gets around the old problem of academics not understanding strategic plans, so even if a firm would license their project, the project would languish on the shelf rather than be developed. Then the company scientists work together with the academics to design experiments for testing how well a drug possibility developed by the academic works with a disease. This collaboration combines academic scientists' deep insights into the particular drug possibility with industry scientists' experience with designing experiments that reflect particular disease conditions. After a specified amount of time, the patent is returned to the university if the tests do not pan out. If they do pan out, new agreements are negotiated.

Finally, the NIH's National Center for Advancing Translational Sciences (NCATS) is also integrating sciences to design the strategic path. For example, they will support broadly applicable rather than disease-specific target-validation approaches and the investigation of non-traditional therapeutic targets that are considered too risky for industry investment. As Collins (2011) outlines, they are developing many new clues such as expanding the types of molecules used as therapeutics, new methods for lead identification, innovations in drug delivery like nano particles, better algorithms for virtual drug design, and better pre-clinical toxicology tests that use 3D tissue engineered organoids representative of human organs like the heart, liver, or kidney. I think that NCATS may lean too much towards improving steps and retaining the old linear and deterministic view, but this concern can be explored.

In the past year I have met other industry scientists who are pushing to expand translational research further by, for example, moving from bench to bedside and back from bedside to bench, and including more of the disease context and patient experiences with the healthcare system. Gittelman (2015) makes a case for a return to clinical research using patients and hospitals, to take into account 'the intact organism' of the diseased human body. Another scientist emphasized the importance of continued networking to integrate, and perhaps move towards imagining configurations of interdependencies among sciences. He is now science director of a company with expertise in interferon assays. Rather than just push his product, he explained how he works with a variety of potential clients to integrate his diagnostic techniques into their strategic path for disease management. In particular, his product can distinguish rheumatoid arthritis from lupus because they have different interferon profiles. Distinguishing the diseases helps to develop new drugs for each. Without that, treatments for both blast away at the immune system in general, which weakens the body's immune response.

All in all, I see a variety of experiments in the knowledge subsystem that are moving towards abductively formulating hypotheses that not only integrate diverse ideas, but imagine configurations of interdependencies to design or redesign the strategic path for drug therapies against particular diseases. Participants in the knowledge subsystem seem to understand that they are generating new clues rather than answers or solutions, and are developing alternatives that have more potential. All these efforts help to take advantage of emergence because they project better pictures of the future for drug therapies that innovators can use to learn about how to move forward.

Evaluating the Imagined Configurations by Elaborating and Narrowing

Finally, or at least finally before cycling through reframing again, the hypotheses that imagine configurations are evaluated. This phase of abductive learning routines needs considerable development because I do not see many examples of systematic evaluation to see if the configurations actually improve the strategic path. The primary thrust of evaluation is to go beyond 'does the configuration work or not?', and instead consider how and why it works or not. The goal is learning, so questions include can we learn more with this configuration of interdependencies about developing possible drugs for schizophrenia, for example, what do we learn, can we make better decisions, can we see further out? Can we surface new consequences, explore better alternatives, and are we learning about the disease in a way that enables us to develop drugs for it? How far do we see, how much better can we develop experiments now, and how well can we explore possibilities more thoroughly? Knowledge system scientists use elaborating and narrowing to evaluate hypothesized configurations of science. Elaborating opens them up to uncovering contingencies and new connections, while narrowing helps to stabilize understandings.

The abductive learning routines generate learning events that emerge when knowledge subsystem scientists think they have learned enough about their configuration to identify the next thrusts in its development.

The learning events are intermediary models that suggest additional implications to be evaluated in ongoing learning. Knowledge subsystem scientists judge if they have arrived at a learning event by reflecting in practice and actively exploring how and how well their configurations work for drug discovery. They seek to take advantage of emergence to see more possibilities, uncover new contingencies, and formulate better hypotheses.

Conclusion

Managing and organizing for innovation depends on building long-term capabilities to support innovation projects. Building long-term capabilities for complex innovation systems requires continually integrating all these sciences and technologies into strategic paths that can guide and enable particular drug discovery innovation projects. The knowledge subsystem addresses the challenges of enhancing the innovation process overall in the complex system of innovation. This subsystem continually creates the means through which project innovators can more and more productively navigate in the labyrinth of new product development. This subsystem creates new understandings of disease processes, better testing regimes and models to evaluate progress and assess safety, more fruitful alternatives for exploration, and other ways to enhance innovation processes—including new ways to create avenues in the labyrinth.

But the knowledge subsystem participants cannot work in the abstract, away from actual innovations. They need to be actively engaged in innovation by working on concrete problems. The knowledge subsystem itself needs to be orchestrated to foster cycling through the abductive learning routines for imagining configurations of interdependencies among particular scientific ideas that may design a strategic path, evaluating that configuration, and reframing it over time. Individual scientists may still propose specific ideas, but the subsystem needs to integrate them. Knowledge subsystem scientists do not need to be discovery scientists, however. They can be actively and concretely engaged if they cycle even partly through the abductive learning routines by publishing studies of interdependencies, not just of separate elements, working at conferences that are devoted to integrating ideas, and participating in alliances with firms in the industry. Scientists already use heedful interrelating around their ideas, so reinforcing heedful interrelating can enable particular drug discovery efforts. I suggested in the introduction that fostering 'T'-shaped skills would facilitate their abilities to think about interdependencies. 'T-shaped' skills means that people are 'not only experts in a specific technical areas, but also intimately acquainted with the potential systemic impact of their particular tasks' (Iansiti 1993: 139).

But creating the knowledge subsystem takes significant transformation in thinking and doing. One transformation is to replace the deterministic idea of prediction with the idea that people need to create the conditions for predictability by constructing, in the case of systems of complex innovation, strategic paths for product innovation. Prediction, however, conforms to the norms of simple rationality which dominate strategic management and institutional management. Another transformation is to rethink the role of science in all disciplines, including the science of science, technology, and innovation studies, from producing novel ideas to integrating novel ideas in ways that at least start down the path of resolving serious societal challenges. Many biology scientists already work on 'use-inspired' challenges that reflect real-world problems, so this transformation affects journals and norms more than it affects everyday science. Publications can report on contemplating intermediary results, reframing problems, and cycles of iterative problem solving through which researchers tinker with technologies and modify categories until ideas and the world seem to match. Treating existing research as intermediary models that are partial and incomplete might prompt the knowledge subsystems throughout academe to build on and rethink actual findings.

Two more big challenges must be addressed before the knowledge subsystem can effectively support innovation. One challenge concerns the need for strategic guidance so that resources can be pulled together to support particular thrusts. The infrastructure cannot innovate in complex innovation systems if project scientists need to stop their product innovation to work on enabling sciences. It also cannot work if knowledge scientists need to stop their integrating work to generate markets—although making their work marketable would be very helpful. Time underlies the strategic problem. The new knowledge from all the new sciences and technologies is immature and emergent in its own right, and will take unpredictable amounts of time to develop. Strategic managers need to stop looking for short-term solutions and start enabling long-term taking advantage of emergence.

The second challenge is to foster the ability to integrate sciences into strategic paths, by developing institutional arrangements that foster collaboration. Pisano (2010) worries that existing institutional arrangements reinforce fragmentation, as everyone races to patent his or her single idea.

A newspaper article relevant to these challenges caught my eye. The article discusses the decline of New Jersey's economy because so many pharmaceuticals are moving to Massachusetts, despite much higher costs in that state. They are moving to be near the many small biotechnology companies there. The writer suggests:

> With patents for blockbuster drugs expiring and little in its pipeline to replace them, the pharmaceutical industry has changed its business model. Why spend billions on time consuming research and development—with no guarantee of a return on your investment—when you can acquire it from smaller companies and universities?

So much for investing in the sciences and technologies needed to support innovation.

5

The Strategic Management Subsystem in the Infrastructure of Complex Innovation Systems

Mustering the Staying Power to Persist and Learn

I closed Chapter 4 with a newspaper quotation stating that some large pharmaceuticals are shutting down internal R&D, and planning to acquire necessary knowledge (for projects and designing strategic paths) from universities and small firms. This is an interesting strategy for a complex innovation system. It presumes that the universities and small firms are developing all this emergent knowledge more productively than in-house R&D, and that they can make the knowledge readily available. It also presumes that large firms can readily absorb all this external knowledge without any absorptive capacity (Cohen and Levinthal 1990). In fact, the strategy requires dynamic capabilities that involve the ability to learn and adapt (Nelson 2005). It also requires long-term partnerships and an intensive degree of managerial attention (Pisano 2006). This strategy requires a rich, well-developed strategic management subsystem.

The strategic management subsystem in the infrastructure for complex innovation systems addresses the problem of continually creating value. As argued in the last chapter, participants in the knowledge subsystem need to break out of the constraining ideology of prediction, and instead deliberately and actively create the conditions for predictability by designing and redesigning the strategic path for drug discovery. Participants in the strategic management subsystem need to break out of the constraining ideology of control by clock-time, and instead deliberately and actively create the conditions for control by mustering the staying power to persist and learn, to borrow a phrase from Lynn et al.'s (1996) study of breakthrough innovations. Mustering the staying power to persist and learn enables the strategic management subsystem to leverage all the knowledge resources that are generated by the first two subsystems already described, as well as by this subsystem. I explain how and why the three abductive learning routines enable the strategic management subsystem to take advantage of emergence by mapping value-creating opportunities richly and deeply into the future.

Managers and scholars of innovation realized that innovation cannot occur without an effective innovation strategy. The literature on innovation management transformed thinking about the strategic management of innovation in the 1980s as well, because strategic managers could not simply emphasize optimizing the current functioning of the enterprise. Technologies and markets emerge and evolve, so value-creating activities must also continually transform qualitatively. Research demonstrated that, to be innovative, organizations require a strategy that defines the goals for innovation, creates the long-term resources to support sustained product innovation, and provides a strategic direction to shape and guide innovative activities (Adams 2004; Brown and Eisenhardt 1998; Leonard-Barton 1995). Focusing on the short-term blocks innovation, because developing the capabilities and business models for innovation takes time. Innovation requires long-term commitment from strategic managers as well as the direction that channels attention and resources. Those of us working on innovation management also believe that all strategies require the capability to generate streams of new products and new processes.

Transforming Strategic Management Subsystem Social Technologies for Complex Innovation Systems

Strategically managing innovation is essential in complex innovation systems too, but complexity introduces three new wrinkles. First, the entire infrastructure needs strategic management for innovation, since knowledge resources are dispersed throughout the infrastructure (Dougherty and Dunne 2011). Recall the polio story in Chapter 1, where leaders of foundations and government units strategically defined the tasks to be accomplished in the infrastructure, and then orchestrated their accomplishment. I use the term 'strategic managers' to refer to all the people in a complex innovation system who work on creating value from all the scientific discoveries. Strategic managers include the managers of the many businesses involved in the

95

innovation system, and also those in many other agencies, public and private, who work on pulling resources together to enhance, commercialize, make better use of, and rethink opportunities. The 'value' in value-creating opportunities ranges from profits to improved public health such as developing and delivering vaccines for malaria.

Second, strategic managers in complex systems have to negotiate their direction in real time by focusing on process rather than content (Stacey 1995). They cannot impose controls a priori, because complex systems react to direction in unpredictable ways (Anderson 1999). Anderson suggests that managers instead establish and modify the direction and boundaries within which innovations emerge by setting constraints on local actions, observing intermediary outcomes, and tuning the process by altering the constraints.

Third, the product cycle time is extremely long, averaging more than thirteen years in pharmaceuticals (very long in other radical innovation as well). The very long product cycle times require that possible opportunities for using new drug possibilities accommodate a larger variety of configurations for new therapies, and be mapped out over the long term. Individual firms develop subsets of possibilities, but together participants in this subsystem would imagine, evaluate, and reframe many possibilities.

Many alternate value-creating approaches have been suggested for this industry so there is no shortage of ideas. Among these ideas are new ways to: (1) develop new classes of therapeutic compounds and treatment modalities; (2) deliver therapies via collaborations with hospitals and patient associations; (3) discover drug families that address diverse genetic makeups among patients; and (4) deploy emerging sciences and technologies such as diagnostics, or research areas like epigenetics into networks of commercial collaborations (Christensen et al. 2009; Pisano 2006; West and Nightingale 2009) However, while there is no shortage of ideas, there may be a shortage of abilities to develop and implement these ideas.

To muster the staying power to persist and learn, managers need to figure out how to explore and implement good opportunities. In this chapter I explain how strategic managers can muster the staying power by using abductive learning routines to stretch strategic horizons beyond the short term. I build on our study of the tensions between managers and scientists over distinct kinds of time pacing (Dougherty et al. 2013a). Time pacing is central to the strategic management of innovation because it is a primary mechanism for guiding, constraining, and tuning local activities. Time pacing coordinates by regulating the intensity and direction of people's attention and efforts (Brown and

Eisenhardt 1998). We find that managers emphasize clock-time pacing, and attempt to pace the development of new drugs 'by the clock', using near-term schedules (Gersick 1994; Brown and Eisenhardt 1997). But drug projects average thirteen years, and are so unpredictable that they cannot be timed by the clock or scheduled by the calendar. Scientists pace their work by anticipated but unpredictable learning events that arise from the cycles of abductive learning. Learning events capture emerging understandings in the innovation, and reflect current and anticipated knowledge resources.

However, we find that clock-time pacing dominates in pharmaceuticals (as it does in many other infrastructures), which hinders long-term thinking. Our study concludes that if innovators and managers can align clock-time pacing and event-time pacing, they can coordinate work further out into the future by imagining more alternatives, and fuller repertoires for responding to opportunities. I extend that idea to argue that enabling both kinds of time pacing allows strategic managers to create the conditions for control by mustering the staying power to persist and learn. Persisting does not mean sticking to a course of action mindlessly. Rather, persisting means continually probing the forward development of a value-creating opportunity and learning about how to improve or redirect its emergence, including choosing to stop developing this opportunity. Aligning clock-time and event-time pacing enables strategic managers to take advantage of emergence because they can coordinate many more activities far longer in time and gauge progress more fully. The strategic abductive learning routines enable managers to align the two time-pacing modalities.

To develop the strategic abductive learning routines that map deeper into the future, I first review arguments from Chapter 1 that the development of new physical technologies like biotechnology require new social technologies. Nelson (2005) and Pisano (2010) emphasize the role of managerial social technologies in industrial emergence, and argue that another major revolution in managerial practice is necessary in the twenty-first century. Then I suggest that the inability to muster the staying power to persist and learn is at the root of current managerial problems with complexity, and illustrate these problems with examples from our research. Industrial society as developed in the nineeenth and twentieth centuries has relied heavily on the social technology of clock-time with the invention of, for example, standard time, railroad schedules, and time and motion studies (Clark 1985; Chandler 1977). Clock-time emphasizes the short term and clear metrics, and does not accommodate alternate temporal structuring. But social technologies that can accommodate more kinds of time have not been developed.

Managerial Social Technologies and Industrial Transformation: Then and Now

I reiterate the discussion of social technologies from Chapter 1, because strategic management is central to the social technologies. Chandler (1977) argues that technological innovation and organizational innovation are interdependent, because new forms of business organization and institutional arrangements are invented to solve specific economic problems (Pisano 2010). For example, technical advances in steam power, steel making, mechanical engineering, and so on may have made railroads and mass production technically feasible, but a host of novel organizational and institutional arrangements made these technical advances economically feasible. These novel arrangements include administrative hierarchies, professional managers, business schools to train those managers, formalized capital budgeting systems, accounting and control systems, and corporate governance structures that separate ownership and management. Chandler had this to say about railroads (1977: 120, quoted in Pisano 2010):

> No other business enterprise up to that time had had to govern a large number of men and offices scattered over wide geographical areas. Management of such enterprises had to have many salaried managers and had to be organized into functional departments and had to have a continuing flow of internal information if it was to operate at all.

Other capital-intensive businesses evolved in a similar way. Advances in the application of mechanical and electric power to production and later to chemicals made mass production technically feasible. But again, without access to capital and the creation of administrative structures to coordinate the diverse activities of these large-scale enterprises, mass production would not have been economically possible. Every organizational form, formal and informal institutional arrangement, principle of management, and management function is the product of human invention, and they all have been developed in response to specific economic problems. According to Pisano (2010: 467): 'After reading Chandler, it is hard to think about technological innovation as anything but tightly intertwined with organizational and institutional innovation.'

Nelson (2005) also says that Chandler's story is about the co-evolution of physical and social technologies. A variety of technological developments during the mid to late nineteenth century opened up the possibility for business firms to be highly productive and profitable if they could organize to operate at large scales of output, and with a relatively wide if connected range of products. Williamson's M form is a social technology that enabled top managers to decentralize yet still control large and diversified bureaucracies. Nelson agrees with Chandler that organization and strategy are as important as investments in R&D, and that key capabilities are dynamic.

The innovation management literature corroborates Nelson's contention that key capabilities are dynamic. Many studies demonstrate that innovative firms continually generate a variety of value-creating opportunities—they rely on economies of scope, not only of scale. Being innovative goes beyond the implementation of specific techniques and procedures, and involves a pervasive orientation that everyone takes responsibility for innovation, not just people in R&D or business development (Van de Ven 1986; Garud et al. 2011a; Dougherty 2001, 2006, 2008). Innovative organizations continually generate new businesses, open new markets, and develop new technologies (Tushman and O'Reilly 1997; Jelinek and Schoonhoven 1990; Brown and Eisenhardt 1997; Danneels 2008; Helfat and Eisenhardt 2004). Innovative firms have the capability to work with different kinds of technologies (Henderson and Cockburn 1996) and build on technological developments made by other firms (Rosenkopf and Nerker 2001).

New Managerial Social Technologies for Twenty-First-Century Industrial Transformation

Pisano (2010: 480) argues that new organizational forms and institutional arrangements are necessary now for science-based businesses like pharmaceuticals:

> Like the railroads and large scale manufacturing enterprises of 100 years ago, science-based businesses will be a potent source of economic growth in the 21st century. And now, as then, these new businesses will demand new organizational forms and new institutional arrangements. In short, we are once again confronted by a serious need to invent new organizational forms and institutional arrangements to deal with a new set of economic problems.

I suggest that the ability to take advantage of emergence over long time periods is an essential aspect of managerial social technologies for the twenty-first century. While innovative firms now engage in a variety of business opportunities, most of the innovations are incremental so firms can rely on clock-time pacing. Brown and Eisenhardt (1997) argue that time pacing, by which they mean clock-time pacing, enhances innovation in software. Change is triggered by the passage of clock-time: a business launches new products every six months without regard for competitive actions, enters new markets every third quarter and not when an opportunity appears, and starts product platforms every twenty-four months.

Despite its benefits for short-term innovations like software products, clock-time makes near-future deadlines most salient (Clark 1985), and shifts attention to exploitation, which undermines exploratory learning (Orlikowski and Yates 2002). Twenty-first-century managerial technologies need to extend the ability to engage in a variety of value-creating opportunities out in time. People can develop longer time horizons if they combine diverse temporal structures to guide, orient, and coordinate their ongoing activities (Bluedorn 2002). Temporal structures are 'expressed in terms of clocks or events, and are created and used by people to give rhythm and form to their everyday work practices' (Orlikowski and Yates 2002: 685). Time, according to Clark (1985: 36), 'is a socially constructed, organizing device by which one set, or trajectory of events is used as a point of reference for understanding, anticipating, and attempting to control other sets of events'. Plural temporal structures enable people to understand, anticipate, and attempt to control a wider variety of events.

The literature on time and organizing captures these diverse temporal structures in two broad categories: *chronos* or clock-time (the serial time of succession measured by the chronometer) and *kairos* or event-time (the subjective living time of invention or people's sense that the time is right) (Garud et al. 2011a; Orlikowski and Yates 2002). However, clock-time or *chronos* dominates modern society and is the taken-for-granted temporal structure that permeates work. In Brown and Eisenhardt's (1997) model, time is clock-time, and does not include other temporal structures. According to Mumford (1936, quoted in Clark 1985: 36), the clock is the most powerful metaphor in the Western world, and has been more influential in the development of capitalism than the steam engine. New time-based social technologies are needed.

Pisano's (2006) summary of ills in the pharmaceutical industry emphasizes the fragmented knowledge that I have detailed in prior chapters, but he also refers indirectly to temporal structuring. He argues that the industry is not structured properly to enable learning for several reasons. First, new small firms monetize their intellectual property (IP), which fragments knowledge. Second, the biopharmaceuticals market for know-how is not based on modularity as it is in telecommunications and other electronic industries, where companies can fairly quickly develop components that fit into the established architecture. In biopharmaceuticals, the knowledge base is immature, companies cannot learn easily or share experience, there are too many subtle interactions, and the technology cannot be codified. Third, funding by venture capitalists is short-term and public finance does not have enough information to value intermediary developments. Pisano (2006) highlights the need to stretch out in time by recommending long-term partnerships to create knowledge, and experimentation with new value-creation models. He also notes that long-term alliances require flexible governance to adapt to changing circumstances, and an intensive degree of managerial attention.

Nelson (2005) does not directly emphasize time either, but he does argue that industrial development progresses more quickly when the technologies involved are largely physical. The relatively slower development of capabilities in the fields of management, education, economics, and health care occur because these fields are dominated by social technologies. Recall this comment that I used in Chapter 1 (Nelson 2005: 208):

> Today, some of our most difficult problems involve developing the social technologies needed to make new physical technologies effective. Arguably the lion's share of the strains in our health care systems are the result of advances in physical and medicinal technologies that societies have not yet learned how to manage or pay for.

In Nelson's view, developing the social technologies is constrained by the tacit nature of the knowledge involved, and by the capabilities, wills, and beliefs of people whose actions must be enlisted, coordinated, or managed.

To put words in Nelson's mouth, participants in these complex innovation systems must allow the needed knowledge to emerge and evolve more fully, which means they need to muster the staying power to persist and learn so they can take advantage of emergence. Extensive analysis is provided in our published papers (Dougherty et al. 2013a; Dunne and Dougherty 2016). For this chapter, I highlight the tendency of strategic managers in this complex innovation system to try to optimize resource use in the short term, to clarify decision-making with more information, and to eliminate uncertainty. I do not suggest that the managers are incompetent, since their job is to generate revenues despite the enormous ambiguity in this industry, and investors demand regular financial reporting. However, they rely on the social technologies developed around clock-time that were created for nineteenth-century industry (Clark 1985; Bluedorn 2002), because they have no other social technologies.

For example, this manager emphasizes getting quick answers to established questions:

> [W]e always have to answer the same questions along the way: is it safe, is it efficacious, does it have the right bio-pharmaceutics properties. But you want to pay less to get to the answer and that is where you use the modern technologies, to find ways to answer the question earlier and cheaper...

He gauges progress by how quickly and cheaply these questions are answered, and seems to gloss over the inherently unpredictable drug discovery process. Another manager also echoes this focus on quick answers:

> The challenge for [the scientists] is how can you answer the question with a greater certainty that you have gotten the right answer and how can you do it more cheaply and how can you do it faster?

In another example, a manager explains that biotechnology companies are more productive because time pressure focuses them on just a few key questions. He thinks that discovery scientists in large pharmaceutical firms like his 'screw around' with interesting questions:

> ... the urgency of having only 2.5 million dollars to get to the end line... You have 2.5 million dollars to answer the question. You don't have time to screw around with eight scientific questions that are interesting but not relevant to answering the product development question.

By implication, he thinks that the relevant questions are obvious, and the answers will arise readily.

Another manager explains that a major problem in their decision-making is the failure to get complete and transparent information:

> What we strive for is completeness of data where possible, but most importantly transparency, because the thing that will be most disruptive to the decision-making is when someone has information and not everyone has information... What you find is the data go all over the place because somebody in the room will know that there was a safety signal and somebody else in the room will know that the chemistry is falling apart and another person in the room will have a perception that this is the biggest drug since [company blockbuster]. You are just all over the place... [The R&D VP] wants to be informed when he makes a decision. He wants to know all of his options and all of his choices, importantly he wants to know what the costs and risks are... Our decision-makers are screaming for this information...

This same manager emphasizes the need for informed and rigorous decision-making:

> We have to really inform our decisions and so we are trying to get to a point where every decision is made with the best possible and most rigorous data in

front of the decision-makers, and that includes things like what is it going to cost, what does it do to the capacity, what are your lost opportunities if you pursue this drug rather than that drug...

The effort to inform decision-making is certainly reasonable. But the emphasis seems to be on objective facts that he feels are readily available.

In one more example, this manager emphasizes reducing uncertainty rather than taking advantage of emergence, and his view of how to go forward focuses on the near term:

> At each step... we create more information and you resolve uncertainty. Uncertainty at the beginning is huge and you are sequentially resolving it. Resolving it could mean I have resolved uncertainty it is a loser or I have resolved uncertainty and it continues to look like a winner... What are the sources of uncertainty? In what way using new technologies and insights can we reduce that uncertainty sooner and can we pay less to resolve uncertainty...

These managers are most likely correct that discovery projects are not as well executed as they could be, but not for the reasons they emphasize. The disappointing R&D performance arises in part from the strategic failure to 'muster the staying power to persist and learn' over many years (Lynn et al. 1996). I do not see much evidence that strategic managers in pharmaceuticals are trying to leverage all the knowledge as effectively as possible, or are providing the strategic framework to shape and guide the innovation work over the long term. For example, outsourcing R&D to small firms and universities could be a good strategy if managers develop the complex relationships needed to take advantage of emergence. But if this strategy is driven by short-term cost cutting, it will fail. I do see evidence that managers are working with a diminished set of events for understanding, anticipating, and attempting to control the future development of new drug therapies. They cannot see very far into the future, and so they cannot see very much.

Abductive Learning about Value Creation to Anticipate a Deeper Future

Strategic managers can break out of the constraining ideology of control by clock-time and create the conditions for ongoing control with strategic abductive learning routines to map out future value-creating opportunities. Recall that I use the term 'strategic managers' to refer to all the people in the complex innovation system who work on creating value from all the scientific and technological discoveries relevant to pharmaceuticals. I argue that the entire system needs a strategic management subsystem that can reach far into the future and take advantage of emergence.

To create this system-wide strategy process, I suggest that participants in the strategic management subsystem use the learning events as clues to possible opportunities. These learning events emerge from the cycles of abductive reasoning in all the subsystems of this infrastructure. Managers use clues from learning events to imagine configurations of interdependencies among strategic resources that might produce a viable new process, business model, collaborative network, and so on. They evaluate these imagined configurations and then reframe them, generating strategic learning events that arise from this cycle of abductive reasoning. The strategic subsystem generates a diverse portfolio of possibilities that emerge and change over time and that inform the entire infrastructure about possibilities. The portfolio affords a deeper look in time because various possibilities work as stepping stones into the future. Some of the emerging value-creating opportunities will create profits, and some will create 'not-for-profit' public health enhancements. The ongoing forward projection of possible value-creating opportunities builds the strategic process for complexity that Anderson (1999) suggests is needed. Continually imagining a portfolio establishes and modifies the direction and boundaries within which innovation emerges.

The strategic abductive learning routines build on the learning events from all subsystems. Learning events are the endogenous occurrences that emerge when scientists and managers learn enough about the configuration of interdependencies they are working on to indicate the next thrust of their innovation work. The cycles of abductive learning routines in all the subsystems produce these intermediary models or rough drafts. Learning events are moments of closure in the exploratory searching that capture enough of the whole configuration of interdependencies to enable people to see what they know so far and to identify plausible next thrusts in their innovation work. Strategic abductive reasoning uses the learning events in two different ways that I will elaborate more fully next.

USING LEARNING EVENTS TO ALIGN TIME PACING

First, managers can align the two temporal structures to muster the staying power to persist and learn. As already summarized, our study revealed two qualitatively different approaches to time pacing: clock-time pacing and event-time pacing. Both clock-time and event-time pacing are temporal—about time—because both mark durations and map out future trajectories by indicating when activities start or stop. Clock-time pacing marks beginnings and ends of activities with clocks and calendars, while event-time pacing marks beginnings and ends of activities with learning events that can be anticipated but when those events might occur is unpredictable. Each perspective on time pacing identifies different trajectories of events and experiences. If they are used together rather than treated as conflicting practices that people must choose between, they can map out more of the future by encompassing more alternatives to be explored.

By focusing on learning events, people in all subsystems can align clock-time pacing with event-time pacing, because learning events reflect both the *chronos* and the *kairos* of drug discovery and development. A learning event reflects the time it has taken to achieve it (*chronos*), and also the subjective sense that an innovation is emerging (*kairos*). We found that scientists seemed comfortable with pacing their work with anticipated but unpredictable learning events. Despite its lack of precision, I think that many of us use event-time pacing as well. For example, Clark (1985) contrasts less experienced business school graduates in a textiles firm with less educated but experienced marketing men with many years of experience at detecting contingent seasonal periodicities in their semi-fashion side of the textile industry. The business school graduates were only experienced with growth, so they mapped out future production based on current processes and missed market changes. The experienced managers were able to anticipate upcoming seasons in terms of styles, colours, textures, raw materials, and the state of the economy, because they relied on a well-developed repertoire of event trajectories which they used to decode equivocality (Weick 1979), from several diverse areas of the total environment. Krupp and Schoemaker (2014) describe business leaders who can look further into the future. Although the authors do not use the term, the managers seem to me to be using event-time pacing to anticipate changes and shift business models.

Event-time pacing regulates the intensity and direction of people's attention and efforts towards cycling through the abductive learning routines. The objective is to get to learning events, which are moments of closure that redirect the work towards the ultimate goal of developing a good drug, a good strategic path, or a good value-creating opportunity. Event-time pacing uses more qualitative metrics to gauge progress. Progress means learning, so metrics involve judgements of how well innovators are learning about the key interdependencies. Innovators and managers can gauge their progress by examining how useful are the configurations they are imagining, if they are deepening their understanding well enough to tackle problems they encounter as they develop their innovation and if not why not, and question how promising are the alternatives and consequences suggested by the imagined configuration. They can also judge if the learning events that are emerging in their trajectory of work are getting richer and more complete. I will detail more indicators in the description of the three strategic learning routines.

By highlighting event-time pacing, people's energy and attention focus on substantive learning, not only on the passage of clock-time. Event-time pacing directs attention and effort to cycling through the abductive learning routines, which leaves clock-time pacing for directing attention and effort to the efficient use of resources for that cycling. Working in tandem with event-time pacing, clock-time pacing creates the ability to generate emerging patterns. Managers and innovators can clock-time pace the development and implementation of the incremental resources that support discovery such as building capabilities in marketing or manufacturing, or developing alliances. Clock-time pacing coordinates ongoing process improvements that will more productively and efficiently support learning in projects and knowledge-integrating efforts. Managers and innovators can also use clock-time pacing to gauge progress by measuring how efficiently people learn to apply new insights, develop supporting infrastructures for particular value-creating models, examine which activities can be done more efficiently to surface interdependencies, identify barriers to learning that can be overcome, and react to results of their experiments. It might even be possible to think about scheduling the launch of new strategic paths and new business models by the calendar, as Brown and Eisenhardt (1997) recommend for software products—provided people understand the complexity of these activities.

USING LEARNING EVENTS TO IMAGINE CONFIGURATIONS OF RESOURCES

The second role played by learning events is to provide clues that fuel the abductive imagining of the strategic configurations of interdependencies. A central task for strategic managers is to figure out what are the essential resources for a given value-creating opportunity and how those resources interact to produce the value. In complex systems, learning events capture the most central resources—what we know so far and think we will learn. These opportunities emerge unpredictably, so the task of developing value-creating opportunities to take advantage of all the scientific breakthroughs is complex and emergent as well. Strategic managers need to delve into how and why particular learning events suggest important resources, how the knowledge signalled by those events depends on other knowledge to produce a possible opportunity, how to acquire and configure all these knowledge resources, and how to implement that bundle effectively.

An example (unfortunately a negative example, which most are) of the strategic interdependencies among knowledge resources helps to illustrate the challenges of building and interweaving scientific, technological, and business operations into value-creating opportunities. McNamee and Ledley (2012) trace the development of three kinds of therapeutic biotechnologies: gene therapy (genetic material administered to transform cells to express a therapeutic gene product), oligonucleotide therapeutics (different technologies to suppress gene expression), and monoclonal antibody therapeutics. All of these were hyped as obvious new therapy approaches. However, monoclonal antibody therapeutics (MATs) are the only biotechnology to be successful so far, because the necessary configuration of knowledge resources to implement this therapy has been developed, and the technology is sufficiently mature to be developed for appropriate clinical applications and markets.

McNamee and Ledley (2012: 943) conclude that successful product development is a 'multidimensional problem that cannot be explained by the advancement or stage of a specific technology'. Rather, it requires the convergence of technologies related to therapeutic entity with clinical practice, understanding of pathogenesis, targets, delivery, manufacturing, bioavailability, markets, and many additional issues. I point out that this convergence is a configuration of knowledge resources that define and direct the whole value-creating process. For example, MATs can be used for chronic diseases as well as acute diseases. But each requires different administrative protocols, new metrics of safety and efficacy, new pricing models, and different distribution channels. McNamee and Ledley (2012) also suggest that biotechnology companies continue to focus on very immature technologies and try to apply them to the very challenging standards of blockbuster products for mature markets. In other words, these biotechnology strategic managers are not imaging, evaluating, or reframing configurations of strategic resources that can actually make use of their technologies.

Cycling through Abductive Learning Routines for Value-Creating Opportunities

A variety of ideas for alternate value-creating opportunities have been proposed, and some are being tried out. I summarize a few of these ideas and then discuss cycling through the three abductive learning routines to develop them more effectively by taking advantage of the emergence of knowledge. We have no shortage of ideas for other complex innovation systems either, like how to fix education systems, how to ameliorate poverty and hunger, or how to overcome ecological damage. The problem in all these systems, including biopharmaceuticals, is to bring possible ideas into fruition so that they actually generate the particular kind of value envisaged, and so that people can continually learn from them to take advantage of emergence.

Along with biotechnologies that are configured with related resources to deliver value, as outlined by McNamee and Ledley (2012), possible value-creating opportunities include building flexible clinical trials, improved distribution and selling capabilities, and enhanced manufacturing. Other new classes of therapeutic compounds such as stem cells, new treatment modalities, and patient specific drugs developed via new diagnostics are also possible. Pisano (2010) recommends new organizational hybrid forms that mix markets and hierarchies, like Genentech. West and Nightingale (2009) suggest new forms like CellCentric, a company that works with more than twenty-five leading international researchers in epigenetics. This company assesses and filters research insights in this domain before they are published. By specializing in a particular area of research (epigenetics), the network can build expertise and stay connected to the world's leading researchers so they can identify innovations with the most potential. Christensen et al. (2009) suggest that small firms can flourish in biopharmaceuticals with improved diagnostics to identify the specific customer base for particular drugs, by serving small markets and currently unserved customers. These ideas require the ability to find, access, and deliver therapies to these small market groups.

Another type of value-creating opportunity concerns strategic transformations of the innovation processes by figuring out what works well and what does not. Scannell et al. (2012) point out that most of the R&D costs are in failures, so they recommend that firms develop a chief dead drug officer to investigate the failures. These failures are learning events, or strings of activities that did not attend to learning events. Together, people in this role across the strategic subsystem would set out the major factors responsible for the progressive decline in R&D productivity and compare different therapeutic areas to explain the differences between them in productivity. Chief dead drug officers can explore the extent to which factors are tractable, such as where what the authors call molecular reductionism and brute force screening are a distraction, and where they do help. Chief dead drug officers can measure the veracity of previous diagnostic forecasting exercises, and examine which clinical test requirements are most costly and least valuable.

Most of these ideas involve fairly intense long-term collaborations among a variety of the firms and agents in the infrastructure, so extensive strategic management with the capability for heedful interrelating is essential. Recall Pisano's (2006) concern that the connections need to be fashioned and refashioned continually in the particular context of the work. As well, certain innovations would have more value to society. For example, the value of a drug that helps with Alzheimer's is much higher than another incremental improvement with existing therapies, even though the profits might be similar. Projecting ahead in this book, the strategic management subsystem depends on innovations in the institutional management subsystem.

Using Learning Events to Imagine Configurations of Resources for Value Creation

The first strategic abductive learning routine uses clues to imagine configurations of interdependencies among resources that would create value. Strategic managers would consider what certain learning events (in projects, knowledge integration, and business/value creation efforts) suggest for a future possibility, and carefully consider how these learning resources work together to create a viable configuration. Going forward in time, managers would hypothesize how the configuration they imagine will emerge based on new current and anticipated learning events, what emerging resource interdependencies might be involved, how the agency or firm can acquire and deploy those resources. I am describing basic business planning, except that this planning reaches out over time, anticipates emergent changes based on learning, and emerges continually.

Strategic managers can perhaps begin the cycle of abductive learning routines by interrogating current business models. For example, what configuration of interdependencies among what resources will generate value in what way by outsourcing R&D? What learning events are this strategy based on, and what are the underlying assumptions that need to be evaluated and reframed? What learning events give credence to the sequential process of drug discovery?

Imagine looking out fifteen to twenty years and seeing a variety of possible value-creating opportunities emerging, depending on various anticipated learning events. The imagined portfolio represents the competitive or strategic landscape out in time. Different firms and agencies would imagine the overall portfolio of possible opportunities out in time, but from their own perspective. For example, the Bill and Melinda Gates Foundation has been working on developing malaria vaccines and other ways to address this debilitating disease in impoverished societies. Their mapping would consider a few different configurations of resources (including not just drug discovery but also marketing and distribution to areas that are difficult to reach), and then hypothesize what each would achieve and how, and what learning events are needed to gauge progress. In this mapping, the Foundation would network with vaccine developments and the various strategic paths that are being designed.

CellCentric (already mentioned, cited by West and Nightingale 2009) is leveraging its foundation of knowledge about epigenetic processes to identify novel drug targets. They are focusing on the most aggressive form of prostate cancer, and developing inhibitors to enzymes that lower the ability of cancer cells to resist current therapeutic agents. They would map out the configuration of resources for developing this particular product along with a potential family of therapeutics, including other applications in other diseases that involve epigenetic related enzymes. They are in alliances with basic research centres, cancer hospitals, and a large pharmaceutical firm. CellCentric looks out into the whole portfolio of emerging learning events from their perspective on epigenetics and prostate cancer.

Which future configurations of resources are possible or plausible depends on what the learning events within a firm and throughout the system are pointing towards, including learning events developed from experimenting with the configuration already. Future configurations also depend on the depth and extent of heedful interrelating that can be developed within and among partners. Ongoing experiments enable strategic managers to figure out what particular knowledge resources depend on to be bundled into an opportunity. Imagining a configuration of interdependencies among resources includes identifying certain assumptions about what would make a configuration a good opportunity, what are people going to learn for value creation by developing the configuration, and how will it help muster the staying power to persist and learn. Different values might include opening a new niche in the market and/or in therapeutics, generating some protection from competition, expanding an existing franchise, and providing a long-term foothold.

Evaluating the Imagined Configurations

The second strategic abductive learning routine evaluates imagined configurations of interdependencies among knowledge resources that constitute new value-creating opportunities. Evaluation assesses whether or not and how the predicted relationships between the configuration of resources and potential opportunities work, and what else the firm or agency can do better. Strategic managers in the infrastructure need to implement and experiment with hypothesized configurations in order to generate evaluative knowledge. By elaborating and narrowing around the interdependencies in the configurations, managers and innovators explore the actual interactions to see how and why their hypothesized configuration actually works. They elaborate out around a subset of interdependencies among resources to consider if these interdependencies are central or not, how and why, and what else they can learn. They narrow in on interdependencies that seem stable and useful, and then elaborate out again to see other possibilities. Elaborating and narrowing bounds and balances new knowledge with existing insights. For example, managers might narrow in on the market and then consider if they can get access to it and how. The goal is to evaluate if this configuration of resources can create value based on why managers thought it would, and to explore underlying assumptions in order to learn.

Some examples of assessing hypothesized value-creating opportunities combine clock-time and event-time pacing. The goal is to make better judgements, not simply better decisions, about why and how this is a good business opportunity. Clock-time pacing questions include: (1) How long does it take us to figure out that we are at a good or bad point? (2) How quickly can we evaluate learning events? (3) How quickly do others provide input to our analyses? (4) How quickly do we identify alternatives and choose among them to take next steps? Event-time pacing evaluative inquiries include: (1) We think we are here, is here good enough for a possible value opportunity? (2) Are we able to handle a larger variety of configurations? (3) How much are we willing to pay to explore potential? (4) Are the learning events that emerge getting better and better? Both temporal structures can address strategic issues such as does this opportunity open a new niche, protect us from competition, extend our existing franchise adequately, give us a long term foothold, and allow us to know more about the opportunity as we also generate revenues?

Reframing the Imagined Configurations

The last phase of abductive learning routines cycles back to the imagined configurations by refining or reframing the interdependencies and resources that are involved. Participants in the strategic subsystem critically examine assumptions, deliberate over different perspectives, and bridge possible differences into new shared directions. The strategic subsystem might refine and replace milestones, and develop new performance objectives that reflect new alternatives and consequences learned from evaluation. The strategic subsystem would also rethink how clock-time and event-time pacing combine, and how well the combination is creating a rich set of reference points that people can use to anticipate more possibilities further into the future. Reframing might revise the future trajectories of anticipated activities that need to be accomplished, and coordinate attention and effort to carry out those activities. Event-time pacing structures the inherently exploratory searching and helps to constrain the short-term nature of clock-time by keeping the future open to emergent possibilities. Clock-time pacing helps to constrain the potentially expansive searching, and marshals the development of resources that can be clocked. The strategic management subsystem identifies new future trajectories and eliminates others based on emergent learning.

This cycle of abductive learning routines takes advantage of emergence when strategic thinkers imagine where they can go with what the infrastructure is learning, and they shape and redirect that learning with ongoing ideas of value creation. They implement imagined configurations of learning events to see how they might work, to see what else seems to be going on, and to surface new possibilities.

Conclusion

Strategies for innovation integrate learning, direct activities, and shape ongoing learning. For systems of complex innovation, strategy making occurs across the infrastructure, as it did for the development of a useful vaccine for polio. However, infrastructures for complex innovation like pharmaceuticals and many others involve multiple cases of polio-like innovation, and so require a portfolio of future possibilities that maps out potentially emerging value-creating opportunities. This future mapping shapes the very long-term development of projects and strategic paths, and of value-creating opportunities themselves. The strategic management subsystem must take advantage of emergence by leveraging available knowledge resources into imagined configurations of interdependencies that can generate value by providing potential applications for emerging project and process innovations. Taking advantage of emergence necessitates the ongoing abductive evaluation of these configurations, using them to learn about possibilities by exploring their implications and seeing what actually emerges. Taking advantage of emergence also necessitates ongoing reframing—including choices to eliminate some possibilities—so that emergent learning can accumulate.

However, it seems that the pharmaceutical infrastructure does not have a well-developed subsystem for strategic management, and the same can be said for most other infrastructures of systems of complex innovation. It seems that managers focus on projects, not on strategic management. We see the same tendency in other infrastructures of systems of complex innovation. Local newspapers recount the failure of local schools to hit proficiency goals but not the failure of the strategic subsystem in education to create better approaches for learning. In the United States, hospitals in the Veteran's Administration continue to take much too long to schedule appointments, but we do not read about the failure of the strategic management subsystem to develop better alternatives for delivery. I am sure that health care agencies in other countries also fail to meet short-term goals and fail to develop improved strategies.

Strategic thinking needs to transform in three ways. First, while individual firms and agencies need sensible strategies, they cannot develop and implement those strategies alone—it takes an infrastructure, since the resources that must be bundled together are dispersed across participants. Many single organizations and their managers do not understand what makes a good innovation strategy, so it is not surprising that the need to generate strategies collectively goes unrecognized. Second, the strategy making for the infrastructure is an ongoing process of taking advantage of emergence, as I summarized in the preceding paragraph. Third, and I think most central, the infrastructure needs to gauge progress in a way that encompasses the very long term, because the development cycles and co-evolutionary emergence take lots of time. But the dominant approach to time management relies on the inherently short-term methods of clocks and calendars.

I have abductively hypothesized a strategizing process that takes advantage of emergence by pulling together the learning events that represent intermediary knowledge in the emergence of particular innovations. Some collaborations across large and small firms and with public research organizations seem to be developing new value-creating opportunities that leverage emerging knowledge, so my hypothesis can be evaluated and reframed empirically going forward.

One assumption in my hypothesis is that the learning events enable participants in the strategic subsystem to pace ongoing developments using event-time, based on the anticipated emergence of learning events in projects, knowledge systems, and value-creating opportunities. This hypothesis in particular needs interrogation through abductive evaluating and reframing, since the ability to imagine further out in time is central for the infrastructure. I suggest that strategic managers can construct a portfolio of possible value-creating opportunities that map out in time, and provide stepping stones into the future. They would continually revise and reframe the portfolio by adding new ideas and dropping others. But the portfolio into the future opens up possibilities for projects and strategic paths, and helps to guide these developments as scientists encounter alternatives and choose options. The portfolio adds additional events that enable people to understand, anticipate, and attempt to control their complex innovations—events that do not exist when only clocks and calendars are used to project the future.

Developing event-time pacing along with clock-time pacing adds additional metrics for gauging progress and making sense of the emergence of projects, knowledge capabilities, and value-creating opportunities. Progress means learning, so metrics involve judgements of how well innovators are learning about the key interdependencies. Innovators and managers can examine the usefulness of the configurations they are imagining for generating new ideas and capturing information, and consider if they are deepening their understanding well enough to tackle problems they encounter in the process of taking advantage of emergence. They can examine the promise of the alternatives and consequences that the imagined configuration assumes, and judge if learning events are getting richer and more complete. Managers can also clock-time how long it takes to determine if learning events have occurred, how quickly people recognize emerging perturbations and evaluate learning events, how quickly others provide input to the evaluating and reframing, and how quickly innovators identify alternatives and choose among them to take next steps.

However, the strategic subsystem cannot take advantage of emergence if the project and knowledge integrating subsystems fail to create products and processes. All the subsystems discussed so far in this book also cannot take advantage of emergence unless the infrastructure generates new governing structures that enable all the collaborations that must take place. I discuss this final subsystem next.

6

The Institutional Subsystem of Complex Innovation

Creating the Collaborative Commons

Infrastructures for complex innovation systems harbour one more problem that must be continually set and solved. This problem involves the institutional challenge of constructing a collaborative commons, where various parties can come together to develop value creating opportunities, strategic paths, or new drug projects. It is the job of the Strategic Management Subsystem, discussed in Chapter 5, to imagine, evaluate, and reframe actual value-creating opportunities by taking advantage of all the learning events. However, strategic managers cannot exhibit the enormous amount of judgement needed to persevere unless they can count on other agencies, both public and private, to share knowledge and engage in long-term partnerships. Project innovators and knowledge subsystem scientists cannot continually generate learning events unless they too can count on others to collaborate with their efforts. The institutional subsystem addresses the problem of continually enabling and then leveraging possible convergences among the multiple trajectories of innovation over the long term.

Hopkins et al. (2007) point out that complementary technologies and organizational innovations in the infrastructure for biopharmaceuticals have taken, and will continue to take, decades to emerge. The emergence depends on a co-evolutionary process in which changes in science, technology, industries, clinical care, and regulations mutually constitute one another. This long time horizon subjects new technologies to large and increasing development costs. Managers, academics, and policy-makers need to promote the incremental, long-term evolution of biotechnology knowledge through a better informed, more effective allocation of public resources. In this chapter, I explain how abductive learning routines enable participants in the institutional sub-system to promote the more effective emergence of technological and scientific capabilities with more effective allocation of public (and private) resources. I propose that institutional participants use abductive learning routines to continually construct several different kinds of 'collaborative commons'.

First, I will define terms, since management academics like me tend to not think about these big things outside the organization, except perhaps as a control factor in regressions. A 'commons' refers to a social system of natural or cultural resources that everyone can access, such as a shared pasture in a village, an ocean, the internet, or associations that focus on sustainable development. A commons is prone to misuse, underinvestment, and free riding, but people do co-construct a commons over time to address complex problems like health care or climate change (Ansari et al. 2013). I focus on how institutional participants imagine configurations of interdependencies among relational elements that enable the collective development of knowledge for innovation among participating organizations.

Second, I label this subsystem 'institutional' because institutional agents create governing structures that enable various parties in complex innovation to work together. Sometimes we think of institutions as if they were mile-wide space ships that hover over us, but Nelson (2005) defines institutions as modes of interaction that enable, not constrain, behaviour. I deepen Nelson's definition with Ansell's (2011) Pragmatist view of institutions that is based on the interpretation and elaboration of meaning about meta-concepts such as 'the court', 'the corporation', or 'curing cancer'. In the Pragmatist view, institutions are dynamic, ongoing interactions between concepts, experience, and situations. So, rather than think of institutional participants as alien space ship soldiers, I think of institutional participants as ordinary people who work with each other to figure out how to generate ongoing connections among their organizations to create knowledge for innovation. For drug discovery, these participants would represent myriad public agencies that fund, develop, and regulate aspects of public health (e.g. NIH, FDA, CDC, Departments of Health at all levels), universities, public research groups and hospitals, not-for-profit patient groups and disease associations, small and large firms, health care delivery systems, and associated professionals.

Third, to define relational elements, the literature on alliances and business ecosystems identifies two categories of relational elements that enable effective learning among participants. One category is a common objective that brings disparate entities together over time, and the second is organizing in a way that enables these entities to continually set goals and resolve conflicts (Dunne 2015). We can elaborate the categories with more particular elements such as leadership, ground rules, and intellectual property regimes to provide an equitable share of assets that are developed.

Finally, on collective development, many kinds of collaborative commons already exist for innovation, and many include public policies, public agencies, and public resources. A short list includes business ecosystems (e.g. 'apps' makers voluntarily work with Android software to generate proprietary components to the benefit of all parties involved), user communities (people freely share ideas for improving or leveraging a proprietary technology), boundary organizations that mediate for profit and not-for-profit development of software, regional networks that enhance the local economy, trading zones where different disciplines can collaborate, and transnational associations that address refugees, climate change, AIDs, and other 'grand challenges' (see summaries by Iansiti and Levien 2004; Wareham et al. 2014; Ferriani et al. 2012; Ansari et al. 2013; Ferraro et al. 2015). Many of these associations rely on taking advantage of emergence to continually construct their collaborative commons, because they collectively transform local bits of noisy information into knowledge that can be used for innovation.

Transforming the Institutional Subsystem Social Technologies for Complex Innovation Systems

Rather than attempt to summarize all the different kinds of collaborative commons that already exist, I focus on two kinds that are central to the evolution of science, technology, and innovation. These are: associations that rely on public resources and governmental agencies, and patterns of technology emergence. I draw from these literatures ideas for how to create and maintain collaborative commons for systems of complex innovation.

Public resources have always played an important role in innovation because much of the knowledge is public, and public agencies work to keep knowledge open and available through publicly funded basic research and intellectual property regimes (Murray and O'Mahony 2007). According to Nelson (2005), the market part of the capitalist engine rests on publicly supported scientific commons. The ambiguity of emergence over such long periods limits investment by private agents in new technologies, so governments typically fund early development, and build and finance extremely expensive infrastructure assets such as roads, gas and water systems, dams, hospitals, and major projects such as rural electrification (Afuah 2003; Tidd and Bessant 2009). According to Dosi et al. (2006), the EU lags behind others in converting science into wealth-generating innovations because EU governments fail to invest enough in basic science, and fail to use industry policies to promote innovative industries.

As well, patterns underlie scientific and technological emergence and structure collective innovation. I want to emphasize that people socially construct these patterns, because social construction plays a big role in the institutional subsystem. The invisible hand of optimal efficiency rarely determines what particular technology is selected to be the dominant design (Tushman and Rosenkopf 1992). We arrive at the IBM PC, the internal combustion engine, or massively scaled electricity generation because of political pressures, market power, emerging learning effects, and often serendipitous dynamics. For example, Microsoft and Intel arose as complementary asset providers for the IBM PC, and dominated the personal computer industry for decades. Patterns in technology evolution include the emergence of dominant designs as already discussed. They also include the emergence of architectures that define interfaces among components and enable many firms to develop components, and the creation of standards and standards-forming bodies to mediate connections among participants in, for example, telecommunications, software, or cloud computing (Dosi 1982; Baldwin and Clark 2000; Piepenbrink 2015). While these patterns are socially constructed, they are also often self-organizing since people work on their own to build components that hook up with the shared core technologies, to coordinate the development of complementary assets, to access suppliers, and to find markets for their products and services.

But complexity introduces two big wrinkles that the institutional subsystem for complex innovation must address in order to generate collaborative commons. One big wrinkle is that complex innovation systems involve considerable public welfare, especially in knowledge-intensive endeavours such as health, education, or environmental issues. And public welfare embodies broad, diffuse objectives rather than clear ends that can be reached with defined means. Public welfare issues are often 'grand challenges' that cannot easily be defined (Ferraro et al. 2015) or analysed (Nelson 2005). People bring varying perspectives and interests to bear on what the outcomes should be and on how to assess progress or effectiveness—all of which cannot be conclusively demonstrated in any case (Latour 2004).

Public welfare also means that public agencies and not-for-profit associations play a significant role in the institutional subsystem of complex innovation systems. However, according to Ansell (2011), when public agencies confront highly differentiated constituencies and adversarial politics—which they always seem to do—unproductive spirals of distrust ensue. Various publics demand greater responsiveness to particular concerns, and if these divergent concerns are not clearly addressed the publics may demand reduced agency discretion. We end up with seemingly incoherent public objectives and rigid bureaucracies that are designed in a way that prevents them from dealing effectively with the complexities they face.

The second big wrinkle is that the structuring from patterns in technology evolution may not work for complex systems. Anderson and Tushman (1990) label the early phase of the emergence of a new technology 'the era of ferment'. In this era, innovators build on different operating principles and so go off in a variety of divergent directions. The era of ferment does not support connections among various parties because knowledge is dispersed around different principles, products are crude and inefficient, and customers cannot figure out what dimensions of value are best. Much of the science and technology in pharmaceuticals and other complex innovation systems are in the era of ferment, and may remain in this era for some time.

These wrinkles indicate the need to construct stabilizing governing structures that can overcome diffuse objectives while assuring public safety and welfare, and overcome the stultifying effects of adversarial politics so that public agencies can play a role. Relations among institutional participants need to persist productively despite the diffuse objectives, respond flexibly over time to emergence, and enable public agencies to take the lead in some collaborative commons. As well, institutional participants may need to jump-start science and technology convergence by deliberately constructing possibilities for integration and evolution.

I envisage several different kinds of long-term collaborative commons that engage different subsets of organizations and agents around different objectives, ranging from the development of pre-competitive strategic paths to the elaboration of certain technologies (e.g. genomics), to the creation of novel sets of clinical trials and delivery systems for particular diseases. To develop this final cycle of abductive learning routines, I draw on the philosophy of Pragmatism as developed by Ansell (2011) to explain how public agencies can work productively on diffuse objectives. Next, I outline four kinds of collaborative commons from the science and technology domain to develop different particular insights for devising and evaluating broad objectives and for organizing to achieve progress. These collaborative commons are:

clusters, persistent innovation cycles, hub and technology platform networks, and outward open innovation. Finally, I suggest a few preliminary ideas for creating collaborative commons for complex innovations systems that build these familiar relational elements into new configurations, and explain how people can use abductive learning routines to imagine, evaluate, and reframe these configurations over time.

Public Institutional Structuring for Collective Innovation

Public agencies actively work in any collaborative commons in complex innovation systems because these systems rely on public resources and embody public safety. Public policy scholars have been grappling with the broad and diffuse objectives of public welfare for some time, so their ideas provide the bones of the collaborative commons, to be fleshed out with ideas from science, technology, and innovation. I draw only on two authors, Richard Nelson and Chris Ansell, so this brief summary is only a beginning. Nelson (2005) defines two problems that complexity generates for public agencies charged with protecting and delivering on public welfare. Ansell (2011) leverages evolutionary learning from Pragmatism to resolve those problems. His discussions of Pragmatism detail how abductive learning routines can operate in the institutional subsystem of innovation, while my abductive learning routines detail how associations among public and private organizations can make progress with such big and diffuse objectives.

Nelson (2005) suggests that public welfare faces the problem of illegitimacy or, put differently, the belief by some that public welfare is best provided by market-governing structures, not by governments. Nelson (2005) reminds us that a market-governing structure is just one of several, and works well when the problem is simple and when only a narrow range of interests is to be accommodated. In market governance, potential users decide how to spend their own money on the demand side and for-profit suppliers provide options on the supply side, all with limited regulations. While people debate the relative degree of non-market governance, all societies rely on non-market-governing structures for various public goods such as education, policing, urban mass transit, rural farm subsidies, and the court system, to mention just a few. An array of public agencies oversee, finance, and guide the ongoing development of pharmaceuticals and health care in most countries as well, because the public has a major stake in this innovation system. Public governing systems arise not because of market failure, but because societies value outcomes like a 'good society', or securing the context for fruitful private lives. Nelson (2005) argues that these objectives go beyond the direct interests of suppliers and buyers.

Nelson (2005) emphasizes one public good that buttresses economic development, the scientific commons. Knowledge is a public good that many can use without using it up, because the fact that one holds an idea does not constrain others from holding it too (Dosi et al. 2006). Keeping science open and publicly funded encourages the competitive exploration of multiple paths that continually open up new possibilities and ongoing evolution. Public science also enables the development of applied sciences like chemical engineering, and of whole new sets of science-based industries (e.g. the invention of computers led to computer science and the invention of transistors led to solid state physics). Nelson (2005) cites a survey of industry managers that shows that they want general results and research techniques from public science to help with problem solving, not specific results that trigger particular projects.

Nelson (2005) worries that pressures to privatize applied science and universities' efforts to monetize their basic research by selling patents may be eroding the scientific commons. He proposes new approaches to patenting and new rules for universities to license widely, because basic science does not produce products, rather it provides the knowledge and tools to wrestle with practical problems more effectively. I discuss patent management only in passing here because many in the pharmaceutical industry think that adjusting the length of patents will solve all the problems. More time under patent protection might help or it might hurt, but it is not a magic solution. It takes an infrastructure of all four subsystems to address systems of complex innovation like pharmaceuticals, along with better intellectual property management.

The second problem faced by public welfare is the organizing challenge arising from the fact that public benefits, including those from the scientific commons, are not easily analysed or organized for. Nelson argues that the greater the number of values and interests that have to come to one collective conclusion before action can be taken, the more cumbersome will be the governance system. He also points out that complex problems that depend more on social technologies are less likely to be effectively addressed.

Ansell (2011) summarizes public policy scholarship that addresses the illegitimacy of public welfare and the distrust of public agencies, the challenges of diffuse objectives, and the potentially cumbersome nature of organizing a non-market-governing structure. With regard to their illegitimacy, Ansell (2011) argues that public agencies are at the nexus of democracy and governance. Public agencies are not the sole source of deliberation, participation, expertise, or problem solving, but they are a unique place for these values. The illegitimacy arises because public agencies may not accomplish what particular constituencies expect them to, so new controls are imposed in an ongoing spiral of distrust and decline. Ansell suggests that we rethink the objectives of public agencies from the immediate execution of narrow functions to ongoing problem solving, and rethink their underlying organizing from hierarchy to heterarchy. Collaborative commons for complex innovation systems also focus on ongoing problem solving and on heterarchical organizing.

On rethinking the objectives of public agencies, Ansell suggests that too often we think of public agencies in accounting terms so we allocate resources to them to achieve narrow functions. However, since their job is to grapple with very broad objectives, they do not achieve immediate functions that can please all of the diverse constituencies. He proposes to replace the notion of narrow functions with evolutionary learning based on Pragmatism (which, by the way, seems a lot like abductive learning routines). Evolutionary learning emphasizes the ability of both individuals and communities to improve their knowledge and problem-solving capacity over time through continuous inquiry, reflection, deliberation, and experimentation. Reflection involves the ability to critically scrutinize one's own common sense and habits, while deliberation builds on reflexive inquiry by exploring different perspectives through ongoing probing, adjudicating, and bridging. Pragmatism emphasizes the provisional, probative, and jointly constructed character of social experimentation, so the results of inquiry are treated as fallible and provisional.

Evolutionary learning redefines the purpose of public agencies from achieving specific functions to engaging continuously yet productively in ongoing processes of problem solving. These processes are directed at ideal ends—what Ansell calls meta-concepts—such as improving education, curing cancer, or enhancing community policing. Pragmatism does not prescribe specific ends or means, and places greater value on open-ended processes of refining values and knowledge. The diffuse objective becomes an ongoing process of working on that objective. Citing Peirce, Ansell argues that meta-concepts entail a soft teleology. The objectives are not concrete future events, rather they are possibilities or ideal end states which the problem-solving process tends toward. Ansell's ideas from Pragmatism resonate well with Stacey's (1995) argument that managers need to focus on the process rather than only on the content of issues, and with Anderson's (1999) argument that managers cannot impose endpoints on complex processes. Managers of complex situations instead establish the direction and boundaries that shape ongoing action, and continually tune the process based on emergent outcomes.

With objectives redefined as ongoing processes of problem solving, the essence of public agencies becomes skill sets and problem-solving capacities that are mobilized for a range of tasks that are not defined ahead of time. Consider, for example, education or health. I do not think that society will ever finally fix education or health problems because these will continually evolve as we learn new ways to address new issues. Ansell notes the irony of current debates over achieving outcomes of the 'common core' in secondary education, to be measured by controversial metrics. Few seem worried about building better skill sets and problem-solving capacities for addressing the complexities of improving education.

Objectives as meta-concepts iterate with actual concrete efforts to solve them, so the objectives cannot remain abstract. Ansell argues that objectives as meta-concepts act as boundary objects that can mean different things to different participants, so people can work together on these objectives without needing to agree on their meaning—they can collaborate without consensus (Star and Griesemer 1989). Different actors have different views about what the problem actually is and what constitutes an acceptable solution (Lindbloom 1959). People work on these different views through concrete action and experimentation, since meaning depends on hands-on experiments that are based on confrontation with concrete problems. Problems disrupt existing assumptions and call for fresh discovery. Problems also pin disputes about knowledge, principles, and values down to particulars. Ansell (2011: 21) concludes:

> Meaning is linked closely to action by focusing on problem solving. Symbols are tools for problem solving, but the uncertainty of problems means that they are treated as provisional (fallible). Problem solving is probative (oriented toward discovery of value through action) and creative (hypothesis forming or abductive).

If the objective of public management is problem solving and the essence of public agencies rests on skill sets and problem-solving capacities, then the hierarchical character of large-scale organizations must transform. Ansell argues that the bureaucratic structure of most public agencies is ill suited for meaningful problem solving and engagement with the public. He proposes that heterarchy replace hierarchy. Hierarchy means one to many links based on hierarchically decomposed subordinate units that are nested within superordinate units, like Russian dolls. Hierarchy presumes a clear ordering of the relationship between superordinate and subordinate units, and that the relationship between the whole and the parts is already well defined. Heterarchy means many to many links among people and units, and resembles a shifting web of multiple lines of communication and interaction. The heterarchial organization is local but holistic, and relies on recursive, continual, and interlocking cycles of perspective. The heterarchy encompasses a network of complementary policies, practices, and strategies that combine to make the public enterprise more effective.

For example, task forces of middle managers from different agencies develop solutions to problems that all agencies face, and hold each agency accountable for implementing these solutions. Or a higher ranked naval officer defers to a lower ranked enlisted officer when the enlisted person has greater technical or contextual knowledge. Weick and Roberts's (1993) heedful interrelating among dispersed units on an aircraft carrier to retrieve aircraft also reflects heterarchy—the ship captain defers to recovery crews on the deck to determine if it is safe to land. Ansell (2011) describes community policing based on local data. Prior to the implementation of the localized approach, the hierarchical policing approach imposed regulations and procedures from the top down to constrain local behaviour, which resulted in organizing to reduce risks and failures. But with the heterarchial approach, local patrol officers meet several times a week with precinct leaders to analyse real-time crime patterns and develop customized crime control strategies. Senior police managers participate in a continuous exchange and mutual adaptation between local problem solving and citywide strategy, while precinct commanders demand resources from top brass and develop shared responsibility. The organization becomes a many linked, recursive cycle among local patrol, specialized units, and strategic managers. Together they actively generate new approaches and ensure that all units cooperate rather than act separately.

Ansell cites a study that defines the deep structure of community policing as accountable autonomy. Neighbourhood efforts must have autonomy to flexibly engage in problem solving, but they must also be accountable to police headquarters. Accountable autonomy seems like a good ground rule for collaborative commons in general.

Although we do not use the term heterarchy, innovation scholars, including myself, describe very similar approaches to organizing for innovation. My model of organizing for innovation emphasizes different sets of experts who take the lead on different kinds of innovation problems, and respond interactively to the activity of others (Dougherty 2006). Rather than issue orders, functional managers develop long-term capabilities to support innovation projects, and make sure that their people contribute effectively to specific project efforts. Project teams build on strategic guidelines but make local decisions, while senior managers develop long-term capabilities and directions. Brown and Eisenhardt (1997) describe semi-autonomous project teams that interact with one another and with senior managers. Jelinek and Schoonhoven (1990) describe three different modes of organizing in innovative firms to address different kinds of issues. Formal organizing defines who has the authority to allocate resources, while quasi-formal organizing pulls together temporary task forces to address new problems such as difficulties moving new chips from design to manufacturing. Informal organizing relies on strong norms for collegial, non-hierarchical interactions among people based on problem solving. Van de Ven et al. (1999), Garud et al. (2011a), and many others add more insights, but most ideas build on many links and recursive cycles among experts who can work autonomously, provided they take responsibility for effective innovation—that is, provided they follow the rule of accountable autonomy.

I summarize Ansell's (2011) evolutionary learning as a process through which public agencies can take advantage of emergence as they grapple with complex objectives. Continual problem solving and experimenting around concrete problem solutions transforms noisy information into useful knowledge. People generate hands-on information as they work on concrete problems, and other people use these experiences to rethink strategies and to guide ongoing problem solving. Heterarchical organizing bolsters the processes of problem solving, as people with different perspectives work on the problem at hand and react to the work of others. Taking advantage of emergence by cycling repeatedly through the abductive learning routines to hypothesize, evaluate, and reframe governing structures for collective action replaces Nelson's cumbersome process with a reasonable one.

However, the bureaucratic and adversarial nature of political life may keep public agencies from implementing evolutionary learning in collaborative commons. I briefly summarize ideas about enabling the co-evolution of innovation and technology, which provide a more concrete understanding of how collaborative commons can actually work.

Private and Not-for-Profit Institutional Structuring for Collective Innovation

The science, technology, and innovation literature identifies different kinds of non-market collaborative commons that enable collective innovation and ongoing co-evolution of technology. I review four of these collaborative commons briefly, to surface details about how organizations with diverse interests can work together on innovation despite diffuse objectives, how they can organize heterarchically, and what might be important ground rules. The first two sets of insights look at collaborative commons from 50,000 feet, and emphasize what they can do more than how they can do it. These ideas suggest conditions that foster ongoing co-evolution. The next two sets of ideas suggest more specific approaches to ongoing problem solving and organizing that enable diverse entities to work together over time.

Clusters

One large literature explores regional configurations or clusters among competitors, industries, public policies, and markets. This literature shows that collaborative commons among diverse organizations and agencies around diffuse objectives are not only possible, they are also prevalent. These organizations operate with non-market-governing structures and seem comfortable doing so. Focusing on clusters for technology co-evolution, firms in these learning clusters outperform other firms by introducing more novel products and being more committed to advancing the technological frontier (Gilbert 2012). Regional agglomerations such as Silicon Valley, the Research Triangle in North Carolina, or the biopharmaceutical clusters around Boston provide participants with access to suppliers and customers, but also to information spillovers that occur because tacit complex knowledge is locally embedded.

I think that clusters help participants turn disparate fragments of information into useful knowledge for creating new products or finding new opportunities, so participating organizations take advantage of the emergence of knowledge. Technology co-evolution and learning are fairly diffuse outcomes that arise from ongoing interaction among the various agencies and organizations in the cluster. Each of the participating organizations most likely focuses on its own purposeful activities but also pays attention to the actions of others in the cluster. It is not as if organizations sit together and draw up a formal agreement to collaborate, although sometimes they might. For most clusters, firms and other organizations participate because the cluster provides considerable value: it enables them to engage more productively in the ongoing but concrete process of solving problems in new product development and commercialization, as Ansell describes.

Gilbert (2012) suggests that certain properties of clusters support or inhibit technological transformation. These properties point to important conditions for effective co-evolution of technologies and knowledge in complex innovation systems. First, properties that support innovation include diverse sources of knowledge and knowledge-creating entities such as universities and major research centres. Second, the industry structure among cluster firms matters, since the more innovative clusters are horizontally structured around the same markets. Gilbert (2012) argues that firms that compete for the same markets can foster radical innovation if they strongly differentiate their products along critical factors that matter to customers. In this setting, competition prompts continued innovation. Cluster firms that are connected by a core technology but operate in different industries can be more innovative as well, because they can leverage knowledge from one industry to another. However, properties that inhibit innovation include vertical clusters, with firms occupying different positions in the value chain, because this structure can reduce the motivation to transition to new technologies. Some structures privilege incumbent firms by supporting existing technology, and push out possible transformations. For example, Sull (2001) suggests that the tyre cluster in Akron, Ohio, supported then current industry operations and blocked local development of a new paradigm that was introduced by an outsider, Michelin (radial tyres, which transformed tyre production and use).

The industry structure opens the cluster to innovation, but innovativeness also depends on the particular relations among firms and on public policies that favour the emergence of new technologies. In an innovative cluster, the network of relations among participants allows the exchange of tacit knowledge and is flexible enough to allow for new connections and new participants. Malmberg and Power (2005) argue that innovative clusters also provide markets for specialized and skilled labour, and enable people to move from firm to firm. With regard to public policies, Gilbert (2012) suggests that policies should favour technology emergence by fostering new startups or providing incentives for new technologies—for example, California's goal of reducing air pollution encourages the development of new automotive technologies. However, public policies such as tax cuts or subsidies stifle innovation if they support an incumbent set of firms and encourage existing technologies to become entrenched in the local economy.

Persistent Cycles of Innovation

Life cycle models provide another broad category of structuring in science, technology, and innovation systems that can foster various connections among public and private organizations. Serghei Floricel argues that innovation systems can persist for long periods of time in each phase of the life cycle rather than move inexorably to the final phase. His ideas fill in some details about the kinds of relationships and public policies that drive continued innovation, and about developing competitive niches for continued technology co-evolution. Some trajectories of technological emergence follow a particular pattern or 'S' curve of evolution, leading to the life cycle model that explains the twentieth-century technology evolution of, for example, automobiles, televisions, or disk drives (Floricel and Dougherty 2007). The life cycle model posits four different stages through which firms move sequentially as they create, develop, exploit, and finally simply use technologies: era of ferment with new technological paradigms, product innovation, process innovation, and incremental innovation.

As I noted before, the era of ferment is considered to be a period of idiosyncratic emergence for new technologies as different operating principles are tried out. But increasing returns to adoption (the more people using the technology, the more valuable it becomes), learning effects, and the development of complementary assets can push industries towards one or a few dominant designs. The development of a particular technology accelerates because people channel their energy into one or a few of them, and customers can now figure out their preferences. The efforts of different firms converge, and innovators build on each other to improve product functioning and reliability in response to various customer needs. Suppliers, buyers, and producers can connect around a common technology, because suppliers develop complementary assets that 'plug in' to the architecture, markets coalesce around different segments of customers who prefer different bundles of features, and producers use learning to drive down costs. Eventually, according to older versions of the life cycle model, firms shift to cost reduction and away from innovation.

However, Floricel observes that biotechnology has persisted for more than thirty years despite the nascent nature of the knowledge, because of vibrant science-technology co-evolution. This science-based cycle requires intellectual property protection, government funds, and a strong scientific ethos. Other science based industries include new materials, new energy systems, and nano-technologies. Innovation systems such as electronics and other digitally based systems persist in the product innovation phase by continually generating new applications and functions, with an exponential growth of performance in core technologies. Examples include telecommunications, semiconductors, computers, and software. And some innovation systems persist in the incremental innovation phase by generating steady cost reductions based on innovations, higher reliability, and new applications for standard capabilities. Examples include the automotive industry, railroads, and electrical generation.

The possible persistence of the science-based cycle in the era of ferment can perhaps overcome the vicious cycling of the era of ferment. Floricel explains that two social processes drive persistence: ongoing knowledge renewal and accumulation, and ongoing funds creation. Knowledge renewal cycles build on the cumulative and combinative potential of the knowledge that is created. However, the historical difficulties in biotechnology suggest to me that institutional participants need to actively socially construct accumulating and combining knowledge, since these processes may not happen otherwise. Institutional participants would need a ground rule of true partnering to create the combinations. A ground rule of true partnering involves appreciating partners' needs and developing one's own technology or science so that it can interconnect with that of others. In her dissertation research, my former student Yun Su (2013) found that some collaborations between basic and clinical research emphasized putting drug possibilities into their 'best shape possible'. This means not only learning about the structures of particular proteins or chemical compounds, but also identifying interactions between the particular drug possibility and human biology or the disease system so that other scientists can use that possibility in drug discovery.

The social process of ongoing funds creation also requires deliberate social construction. Funds renewal depends on a high salience of needs (which exists for health but not for alternate energy or even poverty abatement), a munificent environment, and the ability to create value capture niches that address heterogeneous customer needs. These niches make experimentation possible, and so are essential. Floricel emphasizes the willingness of governments and large firms to continually finance emergent science and technology development, and biotechnology and pharmaceuticals have definitely benefited from extensive public and private financing. However, governmental, industry, and venture capital funding may be eroding in pharmaceuticals.

Funding alone cannot assure co-evolution. Innovators must also work closely with customers to apply their knowledge and learn about particular interdependencies. But many tiers of suppliers and providers separate drug discoverers from patients, including employers who buy health care, insurance companies, hospital systems, and physicians. Effective collaborative commons need to include end-to-end flows between sets of innovators to customers, so that innovators can see how their ideas actually work and leverage those insights.

Two Examples of Focused Collaborative Commons

Two more sets of ideas focus in on the actual workings of clusters for innovation, and detail how participants deal with diffuse objectives and keep participating organizations engaged for the long term. Hubs and platforms build directly on product architectures or templates that define the parts of a system and how they go together, but architectures do not exist in complex innovation systems. However, hubs and platforms point to the importance of the leadership of large firms and industry groups, to the interactive and heterarchical relations among participants, and to processes for intellectual property management.

HUBS AND INDUSTRY PLATFORMS

Nambisan and Sawhney (2011) build on Gawer and Cusumano (2002) and Iansiti and Levien (2004) to describe orchestration processes in network-centric innovation like hubs and industry platforms. While the objective of a hub collaborative commons may seem obvious, it also emerges over time as internal and external technologies change and market needs appear. The objective becomes continuous problem solving as Ansell argues is necessary for dealing with public welfare problems. A hub firm owns the product architecture and the resulting products that are co-generated with partners, but depends on partners to design and develop the innovative components. Boeing, for example, serves as an innovation integrator by defining the basic architecture of an airplane to be created, and inviting network members to design and develop different components. The hub firm provides opportunities for partners to leverage the innovations by offering a common set of technologies and tools that partners can deploy to assure quality and module integration. Everyone in the network can leverage these assets for themselves.

Hub leaders work hard to keep the technology evolving so it does not become stagnant, and also to keep partners engaged. First, leaders like Boeing update the architecture to stay up with or even lead changes in technologies and markets, and to avoid becoming stagnant. Hub leaders envision and champion changes in the network's goals and in the interactions among members. Second, hub leaders manage innovation appropriability by establishing policies and guidelines that assure fair and equitable distribution of intellectual property rights. Third, hub leaders provide partners with easy and transparent access to the hub firm's commercialization infrastructure.

Organizing the hub collaborative commons involves heterarchical relations. Despite their strong position, hub leader firms cannot rely on top-down command and control because they depend on their partners. Boeing depends on its partners to design and build its airplanes, while the partners depend on Boeing to provide a market for their technologies. The hub network uses forums among partners to design and develop assets that everyone can leverage, to redefine partner roles and coordination as architectures emerge, and to develop norms and policies for IP management. For example, to manage the intellectual property issues, Nambisan and Sawhney (2011) suggest that hubs use partner certification practices to enhance trust among partners for sharing assets, involve partners in devising norms and policies related to IP rights management, establish an IP rights committee with partner representation to resolve IP related issues, and implement systems that enhance the transparency of sharing and using IP rights.

An industry platform leader is similar to a hub firm, but the platform leader makes their proprietary technology a platform that enables others to develop complementary assets and to appropriate value from their complementary products. Industry platforms gain control over an installed base, broadly license their IP, and facilitate partner investments in complementary innovation. They also build brand equity and manufacturing, distribution, or service capabilities. Examples of industry platform leaders include Microsoft Windows for computers, Apple iOS for smart phones, and Microsoft's more recent efforts to entice software developers who have defected to Apple or Google back to Windows. In another example, computer firms formed an industry group to create their own open-standard platform for managing the integration of heterogeneous information systems.

OPEN INNOVATION
Fiat's outward-bound open innovation of its R&D assets provides another detailed example of a collaborative commons (Di Minin et al. 2010). In the early 1990s, Fiat was going through troubled times and reduced its support of R&D significantly. To generate income to keep their technologies emerging, the head of Fiat R&D opened up their technology assets in engines, electronic systems, and vehicles to external partners. Fiat outsourced their non-central technologies to companies that were not direct competitors. At one point Fiat R&D had 700 partnerships with external organizations, and generated 70 per cent of their R&D budget from this open innovation. Fiat also participated in EU publicly sponsored research projects which enabled them to carry out free benchmarking with leading research institutes, competitors, and firms from other industries, and share a lot of information on state of the art in promising technologies.

As Di Minin et al. (2010) detail, none of this happened by magic. Fiat R&D dedicated themselves to providing competitive advantage to customers as a matter of principle, and looked for customers who would be good partners for long-term development, based on high mutual co-dependence. They sought customers who would depend on Fiat's technology development and on whom Fiat depended to develop a particular technology effectively. Fiat emphasized learning from concrete developments and experiments with partners. To develop these partnerships, Fiat concentrated on understanding the actual needs of each partner in creating their own value, so the Fiat engineers could figure out exactly how their technologies add value to the partners' value-creating processes.

To execute this customer-focused outward-bound R&D, Fiat developed techniques to help researchers measure and track how their technology contributed to partner/client competitiveness, impacted their business process, and benefited the client's current and future product portfolio, competence, and strategy. Fiat also organized with a very complex and expensive matrix structure that had units dedicated to marketing and communication, and to scouting and scanning in-house technologies that could be transferred to clients in new industries. Each technology area (e.g. engines, electronic systems) assigned people to coordinate external and inter-functional activities. Di Minin et al. say that to make open innovation successful, Fiat needed a committed, visionary, and passionate champion, and the senior executive leadership played a critical role in promoting the transformation of Fiat's innovation status.

To summarize this brief review, non-market-governing structures are prevalent, and they especially enable the co-evolution of technologies, sciences, applications, and regulations. Some collaborative commons work with market governance and some work in addition to market governance. Innovative clusters rely on diverse players and knowledge-creating entities, including many publicly financed universities and research groups. Public agencies and policies that foster new technologies play an active role in innovative clusters. However, competition for the same broad market also helps, provided that participants can generate differentiated niches or share the same technologies across industries. Innovative clusters also depend on active interactions among participants.

The potential for persistent innovation in the era of ferment requires ongoing knowledge renewal and accumulation, and ongoing funds creation (Floricel and Dougherty 2007). I build on the socially constructed nature of patterns in science and technology evolution, and emphasize that people deliberately foster ground rules for ongoing knowledge accumulation, since biotechnology may not have enough self-organizing dynamics for collaboration on its own (Pisano 2006). These ground rules include true partnership, putting biomedical products into the best shape possible (i.e. identifying critical interdependencies so they can be used in drug discovery—Su 2013), and accountable autonomy—accountable to the infrastructure. Persistent innovation in the era of ferment also requires numerous niches for application, so innovators can experiment and learn, and even generate revenues. Institutional participants would need to create and oversee niches for application so that innovators can work with patients, hospitals, and other clinical providers to see how well their proposed technologies work. The profound disconnect between 'products' and users or applications is, I think, one major reason why public and private resources—money and knowledge—have not been effectively allocated or applied. But working with patients involves enormous risks and safety, and so requires carefully developed institutional arrangements to assure this important connection.

The two examples for hubs and Fiat's outward open innovation highlight the active leadership of large firms for creating and sustaining collaborative commons—including creating and sustaining user applications and market access, and heedful interrelating with heterarchical organizing. Hubs and industry platforms rely on strong and active leadership from the central firm or industry group. The leaders enable ongoing collaboration from various suppliers by keeping the underlying technology updated, providing shared tools for building and diagnosing components, and enabling participative forums among partners to resolve conflicts over intellectual property or other issues. Importantly, the hub or platform leader provides a market for the suppliers' products, either directly by buying the components as a hub, or indirectly by building brand equity and other capabilities that keeps customers interested in the platform. Fiat R&D's very close working relationships with partner customers provides a strong example for creating very deep and rich relationships among participants in a collaborative commons. Despite the expense, these relations generated the needed revenues.

However, large or incumbent organizations with public policies that favour existing technology can dominate and squelch co-evolution. Vertical supply chains prompt specific suppliers to look only at their own step rather than at the whole system, and to avoid disruptions that may come from new technologies. The pharmaceutical industry has elements of innovative clustering, but some experts argue that large firms dominate the industry, which can squelch innovation. Others openly advocate for a vertical supply chain in pharmaceuticals to provide everyone a seat at the table, but this structure also can deter innovation.

Abductive Learning Routines

I propose that the institutional participants—those who work on creating collaborative commons for the infrastructure—leverage clues about collaborations that failed or that continue to work to imagine new configurations of interdependencies among relational elements that would generate collaboration for solving particular problems in the project, knowledge, or strategic subsystems. The goal for this subsystem is to devise and continually try out and reconstruct a governing structure that supports and fortifies the development of another innovation. Participants would evaluate those configurations by putting them to use to surface new ideas, examine unexpected contingencies, and scrutinize their own expectations. And they would reframe the configurations of interdependencies among relational elements based on what they learn about the problem they are trying to work on and the approaches that might work.

Recall that Ansell (2011) recommends evolutionary learning based on pragmatism that concentrates on problem solving, reflexivity, and deliberation. Reflexivity refers to scrutinizing one's expectations and perspectives and changing the process to capture learning. Deliberation refers to inquiry based on the clash of different perspectives and communication for probing, adjudicating, and bridging these differences. The abductive learning routines developed in this book map pretty well onto Ansell's ideas for evolutionary learning. I emphasize making sense of interdependencies among elements in a problem space and examining that sense-making by using the imagined configuration to see, or sift through, complex and noisy reality. Ansell emphasizes the deliberative processes that are well developed in political science, and so helps to complement abductive learning routines.

I list the relational elements that this chapter has suggested might be involved in different ways for different collaborative commons. Innovators in the institutional subsystem would hypothesize a configuration of interdependencies among selected elements that would generate a viable collaborative commons for their problem.

- From Ansell, a problem-solving focus to provide critical learning opportunities. The problem for the collaborative commons centres on what people need to collaborate over, and could concern working on a particular new drug therapy, developing improved models to evaluate intermediary drugs or designing other strategic paths for drug development, or trying out a new business model.

- From Ansell, organizing with heterarchy: organizing for participation, inclusion, with different levels working on different aspects of the problem. The Fiat example shows heterarchy, as do the others in various ways.

- From hubs and platforms, self-organizing to some degree, meaning that different firms, entities, and agencies participate voluntarily because they can gain value from that participation. Ferraro et al. (2015) say that what I term collaborative commons require sustained engagement. While continued engagement seems necessary, I do not know if all the same parties need to stay engaged over the long term.

- From all, co-dependence among participants reinforces active participation. Assuring that participants gain new insights and can use any new developments also assures some persistence. Individuals, groups, and organizations must, of course, recognize the usefulness of sharing knowledge and co-developing ideas and innovations. I think that enough people and agencies understand the usefulness of co-development to begin, while those who still believe in going it alone can learn from them.

- Industry structure: Gilbert (2012) warns against vertical structures that assign participants to a separate role in value creating. Competition motivates innovation, if protected niches exist.

- Leadership by individuals and by large firms or coalitions of firms, and by public agencies. Leaders emphasize heedful interrelating by enabling participation, assuring well-functioning processes for deciding on goals and resolving conflicts, and keeping the central 'platform' evolving. Public agencies such as the FDA can successfully lead certain coalitions, provided they can avoid the policing role.

- End-to-end to customers. Innovation needs some protected niches for value creation by individual collaborative commons that can reach customers. Not-for-profit disease foundations, hospitals or hospital coalitions, clinical research groups, or other end users should participate in the innovation process.

- Money. Collaborative commons need funding from participants, governments, large firms, hospital systems, and insurance companies. Money may also be generated by selling intermediary outcomes to interested customer groups.

- Rules of the game: true partnering, heedful interrelating, account-able autonomy.
- Intellectual property protection and development. I will leave this open for others to elaborate, since I know much less about IP and want to emphasize the importance of all other elements for build-ing governing structures (others may concentrate on IP alone, which is a mistake).

The infrastructure for drug discovery and development should start with a few collaborative commons around pressing problems. People can learn from these experiences to develop additional collaborative commons. Experiments are already ongoing, and should provide con-siderable data for formulating, evaluating, and reframing hypotheses about collaborative commons to work on particular problems. The objectives become continuous problem solving rather than the short-term achievement of narrow functions. But the specific problem needs to contain opportunities for active, concrete problem solving. Some intermediary solutions should arise but the problem solving will con-tinue. To develop and cycle through the three abductive learning rou-tines, collaborators need to choose an actual problem of innovation that the collaborative commons will be developed to solve (or work on). Building, evaluating, and revising a collaborative commons over time is an ongoing problem-solving process in its own right, so the institu-tional subsystem deals with two interacting problems.

Cycling through the abductive learning routines starts with imagin-ing a configuration of interdependencies of relational elements that would generate the specific collaboration needed to solve the particular problem. For example, if the problem is to develop improved models for early evaluation of drug possibilities, the collaborative commons should include large firms, public research hospitals, regulators, and also enable small entrepreneurs to participate. Leaders need to be appointed, per-haps with a governing body and smaller task forces to handle particular developments. Who gets what, who pays what, and who bears the risks all need to be worked out, with intellectual property arrangements, rules of the game, and governing arrangements to make particular decisions and to oversee particular processes. The interdependencies articulate what certain relational elements depend on to work. For example, if people think that a certain kind of intellectual property arrangement will draw participants or will assure compliance, they need to specify the connections—which are then evaluated. Leadership by a coalition will require what kind of co-dependence, and what rules for continued participation?

Evaluating digs into the hypothesized interdependencies to check assumptions and to explore how the collaborative commons works. In addition, the collaborative commons requires its own metrics for evaluation, such as how well it addresses conflicts and other unexpected problems, enables engagement, retains participants, and allows new participants to join. Like hub and platform arrangements, the collaborative commons also needs to be able to update and revise the emerging models for evaluating drugs, and the processes for using them. Collaborative commons experiments generate learning events as well, such as achieving a certain degree of engagement and compliance, figuring out how to implement new processes, and jointly developing agreed upon indicators for drug evaluation. Evaluating necessarily includes how well the collaboration addresses the problem it is designed to solve. Reframing the collaborative commons iterates among participants to generate different perspectives and to see about rethinking the problem so that diverse views can be accommodated.

Cycling through the three abductive learning routines should lead to better allocations of public resources—money and knowledge—because people continually assess and learn from assumptions. Specifying expected value also helps to improve the allocation of resources, since participants would evaluate how well progress seems to be occurring, and think about what else they can do to enhance the collaboration. The drug discovery infrastructure has wasted considerable funding and knowledge by assuming that genomics, bioinformatics, microbiology, rational drug design, and other new technologies would just work as promised (Pisano 2006; Scannell et al. 2012). People did not figure out what kinds of collaborations among what organizations and agents in the infrastructure would be needed to enable these sciences and technologies to work—or in fact to co-evolve and emerge—or how to produce these collaborations.

I will mention three brief examples to illustrate possibilities: developing a true platform, a return to clinical research and development (Gittelman 2015), and carrying out the National Center for Advancing Translational Sciences (Collins 2011; discussed in Gittelman 2015). First, people often refer to technologies or sciences in pharmaceuticals as platforms, but they are not. True platforms like Google or Microsoft office are complete systems that provide end-to-end solutions, enable component makers to hook up and generate their own value, and provide market access. One possible collaborative commons could take one of these 'platforms', like capabilities that may be close to application and market access, and work on fleshing it out to be closer to value creation. For example, to become a platform genomics needs diagnostics abilities, access to particular patient groups, hospitals, and medical providers, different sets of drug therapies that fit different genetic make-ups, and ways to create, maintain, and revise the value chain that connects all the parts over time for particular applications. Researchers in health, public policy, and innovation management could collaborate over how to create a platform-like backbone for complex innovations, including whether such a thing would be possible. Meyer and DeTore (1999) develop a service platform based on different components of knowledge, using an insurance business that provides work-related injury recovery as an example. Is it possible to transform these ideas into a collaborative commons for a kind of platform in drug therapies?

A second example comes from Gittelman's (2015) history of clinical research for drug discovery. Clinical research, according to Gittelman (2015), refers to research performed by a scientist and a human subject working together. Clinical researchers carried out drug discovery from the 1940s until the 1970s. The factors of decline included eroding financial support (the rise of managed care reduced funding for lengthy patient stays in hospitals), increased bureaucracy due to pressures for revenue control and administrative guidelines, increased regulation of human subjects, and fewer and fewer young investigators choosing a clinical research career. As well, genomics and other reductionist techniques became an increasingly dominant model for disease research according to Gittelman (2015), and filled the void being left by the decline of clinical research.

However, Gittelman (2015) argues that the productivity paradox in drug discovery—basically that all the new biotechnologies have not improved drug discovery as expected—arose because the logic of reductionist science was applied to a fundamentally complex technological problem: the discovery of effective medicines that work safely in the human body. I add the need to work safely in the diseased human body, because while healthy people might tolerate certain drug side effects, a person with one or more chronic diseases may not. Gittelman (2015) does not think that society can return to the former clinical paradigm because institutional conditions have changed. But she concludes that we need institutional and organizational arrangements that enable technological search to unfold according to its own logic.

I start with Gittelman's conclusion, and suggest that we imagine such an arrangement to revive but revise clinical research, a process that should also include the many calls I have heard from discovery scientists to develop early in-human testing of possible compounds. The factors that led to the decline of clinical research serve as clues for a new institutional arrangement: lack of funding because of shifts to managed care, heavy regulations, large bureaucratic organizations with their own agendas, and few investigators. Experts need to identify a kind of process or disease area that this arrangement should focus on, since concrete problem solving is essential.

A hypothesized configuration for a clinical research collaborative commons would identify participants and leaders, including regulators and large public research groups and hospitals. They should begin with a disease area that will likely lead to outcomes if enough different groups are incorporated. A particular kind of cancer might work, or another disease with an active not-for-profit venture group that already has lined up collaborators. The participants need to figure out financing and revenue sources, which might include some insurance companies who now pay enormous costs for end of life diseases, some public funding, and perhaps service payments for the process from firms. They would also have to figure out how to avoid onerous bureaucratic organizing and form a governing committee to define rules and intellectual property (IP) plans, safety regimes, and funding—all building on the factors outlined that lead to the decline of clinical research. And they would need new regulations for safety and risk that apply specifically to the disease or medical challenge in question.

Abductive learning routines for the institutional subsystem do not focus only on the problem itself, but also on the governing structures that enable people to deal collaboratively with the problem. Participants would evaluate their governing structure for clinical research by focusing on the relational elements: are participants able to resolve conflicts, do we go end to end enough to generate revenues from customers (insurers, hospital systems, patients), can we incorporate new participants over time while retaining initiating participants, is safety adequate, and of course are we effectively addressing the core problem of remediating disease X with clinical research? The objective will be to learn about leveraging clinical studies and bedside to discovery processes to enhance overall drug discovery. Evaluating would include the learning events: do we get to them effectively, can participants now see more, figure out how to do more together? Reframing the collaborative governing structure over time will rework connections based on deliberations among participants.

The NCATS idea would identify the hypothesized interdependencies among relational elements that underlie the new centre's focus on validating leads that academics discover. Gittelman (2015) summarizes others who suggest that this model is flawed because it has assumptions that should be tested. These assumptions include that the problem is a gap to be filled by translating academic into applied research, and that learning flows linearly from bench to bedside. The programme also builds on particular relational elements such as how participants will interact to solve what kinds of specific problems in translation, how they will resolve conflicts, and what IP arrangements will work effectively to protect rights yet enable knowledge sharing.

The public nature of these large experiments in formulating governing structures and then evaluating and reframing also invokes many challenges since public discourse tends to be stunted by adversarial politics that preclude innovation (Ansell 2011). Abductive learning routines provide a methodical, coherent way to work though problems while attending to risks and conflicts. I suggest that using abductive learning routines simultaneously across all four subsystems in the infrastructure of complex innovation systems is the only way to address safety and risk effectively. Effective innovations that solve major problems safely emerge from the entire infrastructure.

Conclusion

The institutional subsystem plays a central and essential role in the infrastructure for complex innovation systems. Without the ability to devise collaborative commons that enable disparate agents and agencies to work together, innovation is just not possible. Collaborative commons provide the ability to socially construct the co-evolution of technologies, sciences, clinical care and applications, regulations, and industry structure—albeit in specific instances of focused and concrete problem solving. I have reviewed just some of the vast literature on non-market collaborative commons that already exist in the domains of science, technology, and innovation to take advantage of emergence. Many of these rely on participation of public agencies and regulators who do not simply police or evaluate as outsiders. Instead, public agencies work with others and revise, reframe, and rethink their particular rules to fit the problem at hand. Together, institutional agents work with others to grapple with broad objectives by focusing on the ongoing but concrete activities necessary to solve particular problems in the larger stream of joint effort.

I have emphasized the idea that institutions are not giant space ships that hover over our heads. People constitute institutions like collaborative commons by working hands-on with each other to create, maintain, and rework their modes of governance, or modes of interactions. DiMaggio (1988: 9) quotes Gouldner (1954: 17) on bureaucracy, and recommends that we substitute institutions. Gouldner says that sociological theories of bureaucracy (substitute common views of institutions or governance) had 'been so completely stripped of people that the impression is unintentionally rendered that there are disembodied forces afoot, able to realize their ambitions apart from human action'. If we think that the 'free market' or some other invisible force is at work, we absolve ourselves from doing the work, and we seriously misrepresent the problem at hand. Ansell's (2011) Pragmatics addresses this sense of disembodied forces afoot in the land. I combine abductive learning routines with Ansell's evolutionary learning to articulate the sensible, observable, everyday actions people can and do collectively take to grapple with institutional problems in complex innovation systems.

The abductive learning routines for formulating, evaluating, and reframing the collaborative commons provides the sensible, observable everyday actions. The collaborative commons itself is an experiment for innovation with no simple outcomes, and so needs continual reflection and deliberation. But reflection and deliberation require something to work on, and that would be the hypothesized configuration of interdependencies among relational elements that people select in the first place. Participants in the institutional subsystem use abductive learning routines to promote the effective emergence of science and technology by enabling co-evolution in particular concrete efforts, and to promote the effective use of public and private resources. One of the central points of deliberation would be the effective use of resources, since the participants would all have some skin in the game.

Another key point is the need to incorporate end users and customers to enable useful experimentation, and to construct niches of actual value creation. Users and customers provide vital insights to innovations and cannot be left out. But in complex innovation systems like pharmaceuticals, end users are hard to access. Project scientists and knowledge scientists cannot access customers on their own. The collaborative commons needs to take on this challenge and incorporate end users, such as companies who want to use the knowledge developments, hospital systems and insurance companies, health care providers, and patient groups. Concerns for safety and risk prompt us to leave out end users. It is not possible to innovate in any subsystem without the active involvement of users, so the collaborative commons all need to address how they will manage safety and risk.

Finally, the infrastructure of complex innovation requires more metrics or ways to gauge progress and effectiveness that go beyond price and costs. The two processes of cycling through the abductive learning routines and pacing with learning events generate additional metrics to guide innovation. The metrics themselves generate controversy and conflict of course, but can include evaluations of how well people hypothesize, evaluate, and reframe their innovation challenges, how effectively they get to learning events and then use them to move on, and how well specific problems in the grand challenges of complex innovation are being surfaced and addressed. Effective collaborative commons can enable societies to focus on especially serious challenges in the infrastructures of complex innovation rather than on the most profitable, or the most easily addressed.

7

Integrating Subsystems into an Infrastructure for Taking Advantage of Emergence

I have proposed a framework for taking advantage of the emergence of knowledge in complex innovation systems. It takes an entire infrastructure to take advantage of emergence in these systems. Taking advantage of emergence refers to grabbing the fragmented and noisy information that abounds in complex systems, integrating or configuring all these bits into potential solutions for actual, concrete problems, and using these configurations to learn about what might work or not, what else seems relevant, and how to reframe understandings to accumulate more noisy bits of information into better and better solutions.

The framework concentrates on the social technologies that empower people to use the burgeoning sciences and technologies to resolve significant social and economic challenges. People create, combine, and recombine scientific and technological knowledge into commercially and socially viable products, procedures, portfolios, and collaborative commons—innovation. Emphasizing the social technologies that enable ongoing innovation across the infrastructure may surprise some people. Our mass media, public and academic, concentrate on the physical sciences and technologies, leaving the impression that new physical artefacts and procedures produce effective uses on their own. They do not.

To highlight the role of social technologies for making use of science and technology, I briefly traced the history of social technologies in modern economies. In the late nineteenth century, social technologies such as large organizations, new accounting, hierarchy, new financial approaches, and new governance systems (among many more) enabled industrialists to stabilize and scale up new physical technologies for steel making, mass production, oil refining, railroads, and so on. In the late twentieth century, new social technologies for managing multifunctional teams and developing new products, turning vertical functions into long-term capabilities, and creating ongoing strategies to identify/create new opportunities enabled people to generate streams of mostly incremental innovations that apply all the new sciences and technologies to produce new functionalities and markets.

Now in the early twenty-first century, sciences and technologies produce considerable new and untapped potential for tackling even more pressing societal challenges. Many of the biomedical sciences that form the foundation for health care and pharmaceuticals are brand new and very much emergent in their own right, while older sciences like chemistry and biology constantly evolve as well, because they cannot rely on reductionist, strictly confirmatory testing. But they are sciences nonetheless. To apply these emerging sciences and technologies to the challenges of complex innovation systems, we need new social technologies for taking advantage of emergence. Three new social technologies organize infrastructures for taking advantage of emergence: (1) a division of labour into four interacting and entangled subsystems that each address a central problem in complex innovation; (2) abductive learning routines that animate the collective processes for figuring things out and creating solutions to problems in all four subsystems; and (3) heedful interrelating for defining and shaping collective roles and relationships.

Product innovators like drug discovery scientists cannot proceed directly to the end of building a functioning and commercially viable product, because they do not know what elements are involved or how to put them together. They must figure out what aspects of science and technology are involved and how these aspects go together as they proceed with the innovation. En route, innovators encounter surprising barriers, new avenues for proceeding that arise unpredictably, and many choices at every turn. But they do not proceed randomly; they navigate through the multi-dimensional labyrinth purposefully, using the tools for navigation they have at hand. Hutchins's study of navigating a large ship (1995; cited by Nelson 2005: 127) emphasizes the cooperative and collective nature of the endeavour. Deciding the direction and steering the ship involve many people with distinct knowledge and skills who are carrying out different tasks. These skills and capabilities are cultural too, since people learn them as they play connected and mutually dependent roles. It also takes an infrastructure to navigate, because some people work collectively on designing and deploying new navigating technologies, others figure out where to navigate to, and others work on the institutional arrangements that allow multiple ships from multiple countries to move through different seas and dock in different ports.

For complex innovation, project innovators navigate through the labyrinth to try and build a viable product. Knowledge system scientists and technologists build strategic paths that define better navigating. Strategic managers create more labyrinths that provide different options, and institutional participants help channel collaborative efforts for all three aspects of navigating. The infrastructure becomes a distributed system for learning and doing that is based on heedful interrelating. I again quote Taylor and Van Every (2000: 207), from Weick's (2005) discussion of imagination and abduction:

> Groups composed of individuals with distributed . . . partial . . . , images of a complex environment can, through interaction, synthetically construct a representation of it that works; one which, in its interactive complexity, outstrips the capacity of any single individual in the network to represent and discriminate events . . . Out of the interconnections, there emerges a representation of the world that none of those involved individually possessed or could possess.

According to Weick and Roberts (1993), the more heed that is reflected in a pattern of interactions, the more developed the collective mind, and the greater the capacity to comprehend unexpected events that evolve rapidly in unexpected ways. With heedful interrelating, people enact aggregate mental processes that are more fully developed than those found in efficiency organizations.

By bracketing out separable challenges to make sure they are fully addressed on their own, building and leveraging abductive learning, and reinforcing heedful interrelating, people can work collectively on complex innovation problems like new drugs in a distributed learning system. Everyone can work on their own aspect of the collective endeavour, and can do so effectively if they know that others are working on related issues that help define their work. Distributed systems already work on drug discovery, just not as effectively as they can. The four subsystems define what people need to work on, and the abductive learning routines and heedful interrelating define how they work. Innovating within each subsystem takes lots of hard work.

First, people in the project subsystem build the product by figuring out what it needs to do and then how they can produce a material entity to do that. They must cycle repeatedly as they go through the labyrinth to figure out what parts are involved, how those parts go together, what else matters, and why. Project innovators attend to the product in action and are deeply engaged in its concrete functioning in the complex domain of use (e.g. the human body, the atmosphere, the inner city). Project innovation transforms from linear sequences of stages and gates to an abductive process of reasoning based on learning events.

Knowledge subsystem innovators integrate emerging sciences and technologies into procedures that design strategic paths for navigating through the labyrinth. These innovations show how to navigate and provide new tools for choosing places to start and making sense of alternatives en route. Knowledge subsystem innovation transforms from focusing on prediction to focusing on creating the conditions for predictability. It also transforms from thinking of the outputs of science and other academic domains as one-off novelties to thinking of these outputs as clues to possible strategic paths.

Strategic management subsystem innovators leverage all the knowledge resources in the infrastructure by using the learning events as clues to value-creating opportunities. They map out an emerging portfolio of possible applications of the innovations that may be emerging. This portfolio guides and shapes the ongoing abductive learning for projects and knowledge subsystem strategic paths. Strategic management works system-wide, as do the first two subsystems. Strategic subsystem innovation transforms from short-term, clock-time pacing to long-term event-time pacing, so the infrastructure can persist and learn far into the future. Strategic managers reinforce event-time pacing along with clock-time pacing to provide richer trajectories of events against which they can evaluate emerging products and programmes.

Institutional subsystem innovators create the ability for the many agents, agencies, organizations, and policy-makers to come together. They generate a variety of new governing structures so that different groups can collaborate on particular strategic, knowledge subsystem, and project problems. Institutional subsystem innovation transforms from only market-based governing structures to include the active, ongoing collaborations among various organizations and agents that enable the co-evolution of sciences, technologies, clinical applications, regulations, and other uses. The objectives become ongoing problem-solving efforts based on skill sets and problem-solving capacities. The collaborative commons generates access to and participation by end users and customers, and actively jump-starts the social construction of co-evolution.

Innovators throughout the infrastructure carry out these innovative activities using abductive learning routines and heedful interrelating. Abduction and heedfulness build on pragmatism, which says that thinking (abductive learning) and doing (heedful interrelating) are two sides of the same coin of the intelligent ability for being in the world.

Meaning is linked with action, and the two mutually constitute each other. Abductive learning is a common form of reasoning based on thinking and doing, and heedful organizing is a common form of organizing based on thinking and doing.

The different chapters apply the new elements in abductive reasoning developed by Dunne and Dougherty (2016) to all the subsystems in complex systems of innovation. Each chapter details how abductive learning works either in general or for particular subsystem problems, so here I summarize a few key ideas. First, the hypothesis to be formulated is a configuration of interdependencies. The configuration captures the whole of the product, strategic path, strategic portfolio, or collaborative commons—all are new and emergent, so innovators cannot assume that any of the 'things' of the situation have already been identified. They must figure out what the things are. Interdependencies capture ambiguities, so by hypothesizing the interdependencies among elements, innovators address the heart of the complex innovation challenge. Rather than build on facts, since there are few if any available, innovators build on clues, which encompass more insight because clues direct people out of perplexity.

Evaluating and reframing are essential aspects of abductive learning, and if they are left out or poorly done, abductive learning will fall apart. Evaluating does not simply test to confirm, although tests can be part of the process. Evaluating digs into underlying mechanisms of the imagined configuration to figure out what they really are and how they work. Innovators elaborate and narrow: elaborating reaches out around interdependencies to explore what else might be involved and to trace out possible consequences, while narrowing delves into particular issues to situate and gather more details. Reframing draws on the perspectives of all the participants to surface different ideas about the emerging configuration, examine conflicts, and adjudicate different assumptions. Reframing depends on collective deliberation from different perspectives and efforts to reconcile diverse ideas into a new configuration of interdependencies.

Heedful interrelating provides a logic of organizing that bolsters the abductive logic of reasoning. To repeat from Chapter 1, heed refers to the way behaviours are assembled—carefully, creatively, purposefully, and vigilantly. With more heed, people interrelate their actions with more care. Heedful interrelating means that people construct their own actions (contributing) while envisioning a system of joint action (representing) and interrelate their action with that of others (subordinating). The actions of one person thus begin to converge with, supplement, assist, and become defined in relation to the imagined requirements of joint action, but only when the joint situation is also represented in the actions of others. People construct and reconstruct their interrelations continually through their ongoing activities of contributing, representing, and subordinating.

Ansell's (2011) discussion of heterarchy replaces hierarchy. Hierarchy means one-to-many links based on hierarchically decomposed subordinate units, and presumes a clearly ordered and well-defined relationship between superordinate and subordinate units. Heterarchy means many-to-many links among people and units, and resembles a shifting web of multiple lines of communication and interaction. The heterarchial organization is local but holistic, and relies on recursive, continual, and interlocking cycles of perspective. The heterarchy encompasses a network of complementary policies, practices, and strategies that combine to make the public enterprise more effective. Many of the analyses of organizing for innovation in our literature describe heterarchies.

I stretch the idea of heedful interrelating in heterarchies to an entire infrastructure, because many knowledge workers already work in this fashion and already engage at least partly in a distributed learning system. Innovators build heedful interrelating into their ongoing abductive learning so that many can participate. People do not need to be fully engaged throughout since some would stay in tune with the activities and track learning events from the outside, and then contribute their pieces heedfully. I showed that existing collaborative commons such as clusters, hubs, and platforms already use heedful interrelating and heterarchical organizing so these ideas are not alien.

Heedful interrelating and heterarchical organizing do, however, conflict with prevailing notions for top-down control and hierarchy. The now old-fashioned social technologies from the nineteenth and twentieth centuries for economic enterprise and innovation management represent the primary barriers to effective heedful interrelating, heterarchical organizing, and abductive learning. We need to get out of our own way. Heedful interrelating does not eliminate active monitoring or assurance of effective performance, rather it enhances these desired outcomes. Organizations still need people in charge of deciding directions and resource allocations, so I am not proposing some sort of headless entity. But these leaders rely on heedful interaction, and on the shifting networks of engaged knowledge workers who come together as needed to work out emerging problems and opportunities.

The proposed framework requires considerable work, of course. Infrastructures for complex innovation systems cannot address their broad objectives unless all four subsystems actively innovate and actively adapt to the activities in the other subsystems. However, how the strategic management and institutional subsystems can actively innovate is the least well developed in the framework, and I think in our literature overall. I proposed the collective development of a portfolio of value-creating opportunities to strategically shape the ongoing innovation across an infrastructure, but how this portfolio can be generated and how it might help provide stepping stones into the future for particular innovation products and programmes needs more research. Strategic management depends on the ability to come together and work out modes of interaction to share knowledge, pool ideas, and provide participants with an equitable share of assets. We see strategic management and new governance arrangements in major social achievements like the polio story (Yaqub and Nightingale 2012). The question remains whether societies can generate a few alternate collaborative commons that expand the scale and scope of these big achievements into ongoing problem solving. My framework does not take into account the power and politics that others address well, so this area needs much more development.

In addition to examining the framework itself, research can also consider how it might apply to other complex innovation systems. Unfortunately because of space, time, and knowledge constraints, I have primarily used examples from the complex innovation system of pharmaceuticals, which may not interest everyone. Many people cited here apply aspects of these ideas to other infrastructures of complex innovation systems, and can be starting off places for further development. For example, Ansell (2011) summarizes many efforts to work on complex problems in the public domain, while others develop more possibilities (see e.g. Ansari et al. 2013; Nembhard et al. 2009; Piepenbrink 2015).

The ideas developed in this book apply to other infrastructures of complex innovation, although adjustments are no doubt necessary. Consider, for example, alternative energy. Innovation projects would include smart grid technologies, or new kinds of biofuels, or fuel cells. Many specific projects already exist and innovators have developed working models, but these innovations do not yet diffuse widely because they cost too much relative to fossil fuels, and I think because there is no infrastructure to make them useful. Alternative energy people I met in a focus group said their biggest problem was a lack of customers. They do not have a market, I think, because their innovations do not fit together into a new system, and do not fit into the old either. We can ask how well the alternate energy knowledge subsystem integrates different sciences into functioning platforms that comprise working strategic paths for project innovators. I suspect not well. We can also ask how well the strategic management subsystem generates new value-creating opportunities to provide the projects with markets and applications. The very large organizations that own and operate the distribution and generation grids would need to actively participate in significant innovation, but they most likely do not. I am not aware of any collaborative commons.

Four concluding comments map the path forward: the hard work needed to actually generate useful innovations, safety and risk management, better use of society's resources, and the modest but essential contributions from innovation management that enable us to use what we already know, and get on with it.

First, hard work: product innovators in the 1980s and 1990s had to learn the hard way that just sticking cool new technologies out into the market and expecting the world to beat a path to their door did not work. It is a lot of fun to search through databases of cool new technologies that never became products or useful processes, and ask what were these people thinking? Innovation does not work that way. Innovation by definition combines, integrates, and builds together all the elements that constitute the functioning of that innovation in the real world of application. Each subsystem of innovation in the infrastructure needs to carry out this same hard work: identify the elements that may be involved, figure out how they might go together, centre on the interdependencies among the elements to capture ambiguities, and try them out repeatedly to see what else needs to be included, and how to rethink the configuration. The biotechnology and pharmaceutical worlds use the term 'platform' to characterize a variety of technologies in, for example, genomics, diagnostics, or bioinformatics. These are not platforms. A platform, like the Boeing architecture or the Google system, encompasses an entire system that actually works in the real world and enables others to work with it to develop particular new applications. Cool new technologies that might have some application if only someone would pay a lot for them and figure out how to use them are not platforms. They are fun entries into the database of crazy ideas that people did not convert into true innovations.

Second, not much has been said about safety and risk management, even though both are central aspects of complex innovation systems. Leveson et al. (2009) point out that safety is an emergent property of the entire system. People cannot impose safety from the top-down with generic rules, and they cannot build safety in from the bottom-up by assuring all the components are properly built. A nuclear power plant is not safe because all the valves work. The relations between a valve and other plant components determine safety.

The same infrastructure-wide source of safety exists for complex innovation systems. It takes an infrastructure to assure safety because innovators need system-level information about the situation in order to make safe decisions. In the case of drugs, most safety regulations concentrate on projects with generic rules about toxicity and clinical trials. But project innovators cannot address all safety concerns because of the complexity. Better knowledge system integrations that assess interactions or enable particular evaluations would provide more safety. Better value-creating opportunities that combine particular delivery programmes that address unique safety issues in patient sub-populations provide more safety again. And active collaborative commons can address the difficult trade-offs of trying to determine how much risk exists in particular activities and decisions, how much risk is acceptable, and how to achieve multiple system goals, which adds even more safety.

Effective risk management is also an emergent property of the entire system and occurs in the infrastructure. Abductive learning routines dig into risks by embracing possibilities for unexpected outcomes and unknown contingencies. If innovators in all subsystems engage in their part of a particular innovation process, they put more eyes on the emerging development, and enhance the likelihood that people will spot even minor perturbations that might signal major concerns. Scientists generate the conditions for predictability more fully if all subsystems are working out their particular problems at the same time. Strategic managers generate the conditions for mustering the staying power to persist and learn only if the other subsystems are working effectively on their problems. Institutional participants assess risky trade-offs only if the others are generating ideas and insights.

An infrastructure based on abductive learning and heedful interrelating addresses many sources of unacceptable risk and also addresses sources of wasted resources. These wasteful but risky activities include implementing immature technologies without thinking through the interdependencies that would make them work, relying on poorly conceived goals and agendas, closing projects or moving them forward prematurely, searching for facts when none exist, relying on clock-time for unpredictable learning events, ignoring our abilities for developing a collective sense of readiness, working towards a very long-term future without imagining possibilities, imposing clear criteria on equivocal situations, and throwing money at problems rather than digging in to continually set and solve them with concrete actions that can be evaluated and revised. We do not need to throw more money at some of our complex innovation systems (although others might need more thoughtful public investment). We do need to leverage the possibilities for discovery and learning.

Finally, this book builds primarily on ideas from science, technology, and innovation management. Many other literatures contribute similar ideas and provide similar kinds of overviews. However, innovation management makes modest but essential contributions because it provides specific details about the everyday, doable, and observable activities and social practices that constitute ongoing innovation. Let's get on with it. The framework developed in this book is one large abductive hypothesis that my colleagues and I have reasoned through with just a few cycles of abductive learning routines. These ideas require much more development, with concrete applications to actual events for resolving problems in pharmaceuticals, health care, and other complex innovation systems, as such as restoring damaged ecosystems, revitalizing inner cities, overcoming the recent crisis in policing and racism in the US and elsewhere, or public education at particular levels.

However, many of the ideas in this framework for taking advantage of emergence have already been developed at some length by many other people. We already know about most of these possibilities. It is time to go try them out, research how and why some configurations work for some subsystems while others do not, explore new options, put ideas into action, and learn from these efforts to tackle complex innovation systems. Taking advantage of emergence is one central way for developing the knowledge needed to resolve important social and economic challenges. Let's take advantage of emergence.

Bibliography

Adams, M. 2004. *PDMA Comparative Performance Assessment Study.* Chicago: PDMA Foundation.

Afuah, A. 2003. *Innovation Management: Strategies, Implementation, and Profits.* New York: Oxford University Press.

Anderson, P. 1999. 'Complexity Theory and Organization Science'. *Organization Science*, 10: 216.

Anderson, P., and M. Tushman. 1990. 'Technological Discontinuities and Dominant Designs: A Cyclical Model of Technological Change'. *Administrative Science Quarterly*, 35: 604.

Ansari, S., F. Wijen, and B. Gray. 2013. 'Constructing a Climate Change Logic: An Institutional Perspective on the "Tragedy of the Commons"'. *Organization Science*, 24: 1014.

Ansell, C. 2011. *Pragmatist Democracy: Evolutionary Learning as Public Philosophy.* Oxford: Oxford University Press.

Arora, A., and A. Gambardella. 1994. 'The Changing Technology of Technological Change: General and Abstract Knowledge and the Division of Innovative Labour'. *Research Policy*, 23: 523.

Arthur, B. 2014. *Complexity and the Economy.* Oxford: Oxford University Press.

Bacon, G., S. Beckman, D. Mowery, and E. Wilson, E. 1994. 'Managing Product Definition in High Technology Industries: A Pilot Study'. *California Management Review*, 36: 34.

Baldwin, C., and K. Clark. 2000. *Design Rules.* Cambridge, MA: MIT Press.

Barley, S. 1996. 'Technicians in the Workplace: Ethnographic Evidence for Bringing Work into Organization Studies'. *Administrative Science Quarterly*, 41: 404.

Barry, A. 2005. 'Pharmaceutical Matters: The Invention of Informed Materials'. *Theory, Culture and Society*, 22: 51.

Bensaude-Vincent, B., and I. Stengers. 1996. *A History of Chemistry.* Cambridge, MA: Harvard University Press.

Bluedorn, A. C. 2002. *The Human Organization of Time: Temporal Realities and Experience.* Stanford, CA: Stanford University Press.

Brown, S., and K. Eisenhardt. 1997. 'The Art of Continuous Change: Linking Complexity Theory and Time-Paced Evolution in Relentlessly Shifting Organizations'. *Administrative Science Quarterly*, 42: 1.

Brown, S., and K. Eisenhardt. 1998. *Competing on the Edge: Strategy as Structured Chaos.* Boston, MA: Harvard Business School Press.

Burns, L. R. 2005. *The Business of Healthcare Innovation*. New York: Cambridge University Press.

Carter, R. 1965. *Breakthrough: The Saga of Jonas Salk*. New York: Trident Press.

Chandler, A. 1977. *The Visible Hand: The Managerial Revolution in American Business*. Cambridge, MA: Harvard/Belknap.

Chesbrough, H. 2003. 'The Era of Open Research'. *Sloan Management Review,* 44: 35.

Chiles, T., A. Meyer, and T. Hench. 2004. 'Organizational Emergence: The Origin and Transformation of Branson, Missouri's Musical Theaters'. *Organization Science,* 15: 499.

Christensen, C., J. Grossman, and J. Hwang. 2009. *The Innovator's Prescription: A Disruptive Solution for Health Care*. New York: McGraw-Hill.

Clark, K., and T. Fujimoto. 1991. *Product Development Performance*. Boston, MA: Harvard Business School Press.

Clark, P. 1985. 'A Review of the Theories of Time and Structure for Organizational Sociology'. *Research in the Sociology of Organizations,* 4: 35.

Cohen, J. 2011. 'The Human Genome, a Decade Later'. *Technology Review* (Jan./Feb.): 40.

Cohen, M. 2007. 'Reading Dewey: Reflections on the Study of Routine'. *Organization Studies,* 28: 773.

Cohen, W., and D. Levinthal. 1990. 'Absorptive Capacity: A New Perspective on Learning and Innovation'. *Administrative Science Quarterly,* 35: 128.

Colapietro, V. 2009. 'A Revised Portrait of Human Agency: A Critical Engagement with Hans Joas's Creative Appropriation of the Pragmatic Approach'. *European Journal of Pragmatism and American Philosophy,* 1: 1–23.

Collins, F. 2011. 'Reengineering Translational Science: The Time is Right'. *Science Translational Medicine,* 3: 1.

Cooper, R. 1998. *Product Leadership: Creating and Launching Superior New Products*. Reading, MA: Perseus Books.

Corley, K., and D. Gioia. 2011. 'Building Theory about Theory Building: What Constitutes a Theoretical Contribution?' *Academy of Management Review,* 36: 12–32.

Cusumano, M., and K. Nobeoka. 1998. *Thinking Beyond Lean*. Boston, MA: Harvard Business School Press.

Dane, E., and M. Pratt. 2007. 'Exploring Intuition and its Role in Managerial Decision Making'. *Academy of Management Review,* 32: 33.

Danneels, E. 2008. 'Organizational Antecedents of Second-Order Competences'. *Strategic Management Journal,* 28: 519.

Day, G. 1990. *Market Driven Strategy: Processes for Creating Value*. New York: Free Press.

Denrell, J., C. Fang, and D. Levinthal, D. 2004. 'From T-Mazes to Labyrinths: Learning from Model-Based Feedback'. *Management Science,* 50: 1366.

DiMaggio, P. 1988. 'Interest and Agency in Institutional Theory', in Lynn Zucker (ed.), *Institutional Patterns and Organizations,* 3–23. Cambridge, MA: Ballinger Publishing Co.

Di Minin, A., F. Frattini, and A. Piccaluga. 2010. 'Fiat: Open Innovation in a Downturn (1993–2003)'. *California Management Review,* 52: 132.

Dosi, G. 1982. 'Technological Paradigms and Technological Trajectories'. *Research Policy*, 11: 147.

Dosi, G., P. Llerena, and M Labini. 2006. 'The Relationship between Science, Technologies and their Industrial Exploitation: An Illustration through the Myths and Realities of the So-Called "European Paradox"'. *Research Policy*, 35: 1450.

Dougherty, D. 1992. 'Interpretive Barriers to Successful Product Innovation in Large Firms'. *Organization Science*, 3: 179.

Dougherty, D. 2001. 'Re-Imagining the Differentiation and Integration of Work for Sustained Product Innovation'. *Organization Science*, 12: 612.

Dougherty, D. 2006. 'Organizing for Innovation in the 21st Century', in Stewart Clegg, Cynthia Hardy, Thomas Lawrence, and Walter Nord (eds), *Handbook of Organization Studies*, 2nd edn, 598–617. London: Sage.

Dougherty, D. 2007. 'Trapped in the 20th Century? Why Models of Organizational Learning, Knowledge, and Capabilities Do Not Fit Bio-Pharmaceuticals, and What to Do about That'. *Managerial Learning*, 38: 265.

Dougherty, D. 2008. 'Bridging Social Constraint vs. Social Action to Design Organizations for Innovation'. *Organization Studies*, 29: 271.

Dougherty, D. 2015. 'Taking Advantage of Emergence', in Raghu Garud, Barbara Simpson, Ann Langley, and Haridimos Tsoukas (eds), *The Emergence of Novelty in Organizations*, 157–79. Oxford: Oxford University Press.

Dougherty, D., and D. Dunne. 2011. 'Organizing Ecologies of Exploratory Innovation'. *Organization Science*, 22: 1214.

Dougherty, D., and D. Dunne. 2012. 'Digital Science and Knowledge Boundaries in Complex Innovation'. *Organization Science*, 23: 1467.

Dougherty, D., and C. Hardy. 1996. 'Sustained Product Innovation in Large, Mature Organizations: Overcoming Innovation to Organization Problems'. *Academy of Management Journal*, 39: 1120.

Dougherty, D., and T. Heller. 1994. The Illegitimacy of Successful Product Innovation in Established Firms'. *Organization Science*, 5: 200.

Dougherty, D., and H. Takas. 2004. 'Heedful Interrelating in Innovative Organizations: Team Play as the Boundary for Work and Strategy'. *Long Range Planning*, 37: 569.

Dougherty, D., H. Barnard, and D. Dunne. 2005. 'The Rules and Resources that Generate the Dynamic Capability for Sustained Product Innovation', in Kimberly Elsbach (ed.), *Qualitative Organizational Research*, 37–74. Greenwich, CT: Information Age Publishing.

Dougherty, D., H. Bertels, K. Chung, D. Dunne, and J. Kraemer. 2013a. 'Whose Time is it? Understanding Clock-Time Pacing and Event-Time Pacing in Complex Innovations'. *Management and Organization Review*, 9: 233.

Dougherty, D., D. Dunne, and E. De Lia. 2013b. 'Organizing for Complex Innovation', in Ben Kedia and Subbash Jain (ed.), *Restoring America's Global Competitiveness through Innovation*, 28–55. Cheltenham: Edward Elgar Publishing.

Dunne, D. 2015. *The Capabilities that Drive Inter-Organizational Learning in Complex Innovation*. Working Paper. New York: Fordham University.

Dunne, D., and D. Dougherty. 2016. Abductive Reasoning: How Innovators Navigate in the Labyrinth of Complex Product Innovation, *Organization Studies*, forthcoming.

Engell, J. 1981. *The Creative Imagination: Enlightenment to Romanticism*. Cambridge, MA: Harvard University Press.

Feldman, M., and B. Pentland. 2003. 'Reconceptualizing Organizational Routines as a Source of Flexibility and Change'. *Administrative Science Quarterly*, 48: 94.

Ferraro, F., D. Etzion, and J. Gehman. 2015. 'Tackling Grand Challenges Pragmatically: Robust Action Revisited'. *Organization Studies*, 36: 363.

Ferriani, S., F. Fonti, and R. Corrado. 2012. 'The Social and Economic Bases of Network Multiplexity: Exploring the Emergence of Multiplex Ties'. *Strategic Organization*, 11: 7.

Fleming, L., and O. Sorenson. 2004. 'Science as a Map in Technological Search'. *Strategic Management Journal*, 25: 909.

Floricel, S., and D. Dougherty. 2007. 'Where do Games of Innovation Come from? Explaining the Persistence of Dynamic Innovation Patterns'. *International Journal of Innovation Management*, 11: 65.

Foss, N., K. Laursen, and T. Pedersen. 2011. 'Linking Customer Interaction and Innovation: The Mediating Role of New Organizational Practice'. *Organization Science*, 22: 980.

Frye, S., M. Crosby, T. Edwards, and R. Juliano. 2011. 'U.S. Academic Drug Discovery'. *Nature Reviews: Drug Discovery*, 10: 409.

Garud, R., Gehma, J., and Kumaraswamy, A. 2011a. 'Complexity Arrangements for Sustained Innovation: Lessons from 3M Corporation'. *Organization Studies*, 32: 737.

Garud, R., C. Bartel, and R. Dunbar. 2011b. 'Dealing with Unusual Experiences: A Narrative Perspective on Organizational Learning'. *Organization Science*, 3: 587.

Gavetti, G., and Levinthal, D. 2000. 'Looking Forward and Looking Backward: Cognitive and Experiential Search'. *Administrative Sciences Quarterly*, 45: 113.

Gawer, A., and M. Cusumano. 2002. *Platform Leadership: How Intel, Microsoft, and Cisco Drive Industry Innovation*. Boston MA: Harvard University Press.

Gersick, C. 1994. 'Pacing Strategic Change: The Case of a New Venture'. *Academy of Management Journal*, 37: 9.

Gibson, C., and J. Birkenshaw. 2004. 'The Antecendents, Consequences, and Mediating Role of Organizational Ambidexterity'. *Academy of Management Journal*, 47: 209.

Giddens, A. 1979. *Central Problems in Social Theory: Action, Structure, and Contradiction in Social Analysis*. Berkeley, CA: University of California Press.

Gilbert, B. 2012. 'Creative Destruction: Identifying its Geographic Origins'. *Research Policy*, 41: 734–42.

Ginzberg, C. 1988. 'Morelli, Freud, and Sherlock Holmes: Clues and Scientific Method', in U. Eco and T. Sebeok (eds), *The Sign of Three: Dupin, Holmes, Peirce*. Bloomington, IN: University of Indiana Press.

Gittelman, M. 2015. *The Revolution Re-visited: Clinical and Genomics Research Paradigms and the Productivity Paradox in Drug Discovery.* Working Paper. Newark, NJ: Rutgers University, Rutgers Business School.

Gouldner, A. 1954. *Patterns of Industrial Bureaucracy.* Glencoe, IL: Free Press.

Grandori, A. 2010. 'A Rational Heuristic Model of Economic Decision Making'. *Rationality and Society,* 22: 477.

Grinnell, F. 2009. *The Everyday Practice of Science.* New York: Oxford University Press.

Hacking, I. 1983. *Representing and Intervening.* Cambridge: Cambridge University Press.

Hall, S. 2014. 'Neuroscience's New Toolbox'. *MIT Technology Review,* 117 (July/Aug.): 20.

Hanson, N. 1958. *Patterns of Discovery: An Enquiry into the Conceptual Foundation of Science.* Cambridge: Cambridge University Press.

Harrowitz, N. 1988. 'The Body and the Detective Model: Charles Peirce and Edgar Allen Poe', in U. Eco and T. A Sebeok (eds), *The Sign of Three, Dupin, Holmes, Peirce.* Bloomington, IN: University of Indiana Press.

Hayek, F. 1988. *The Fatal Conceit.* London: Routledge.

Helfat, C., and K. Eisenhardt. 2004. 'Inter-Temporal Economies of Scope, Organizational Modularity, and the Dynamics of Diversification'. *Strategic Management Journal,* 25: 1217.

Henderson, R., and I. Cockburn. 1996. 'Scale, Scope, and Spillovers: Determinants of Research Productivity in Drug Discovery'. *Rand Journal of Economics,* 27. 32.

Henderson, R., Orsenigo, L., and Pisano, G. 1999. 'The Pharmaceutical Industry and the Revolution in Molecular Biology: Interaction among Scientific, Institutional, and Organizational Change', in David Mowery and Richard Nelson (eds), *Sources of Industrial Leadership.* Cambridge: Cambridge University Press.

Hopkins, M., P. Martin, P. Nightingale, A. Kraft, and S. Madhi. 2007. 'The Myth of the Biotech Revolution: An Assessment of Technological, Clinical and Organisational Change'. *Research Policy,* 36: 566.

Hutchins, E. 1995. *Cognition in the Wild.* Cambridge, MA: MIT Press.

Iansiti, M. 1993. 'Real-World R&D: Jumping the Product Generation Gap'. *Harvard Business Review* (May–June): 138.

Iansiti, M., and R. Levien. 2004. *The Keystone Advantage.* Boston, MA: Harvard Business School Press.

Jelinek, M., and C. Schoonhoven. 1990. *The Innovation Marathon: Lessons from High Technology Firms.* Oxford: Basil Blackwell.

Joas, H. 1996. *The Creativity of Action,* tr. Jeremy Gaines and Paul Keast. Cambridge, MA: Harvard University Press; orig. publ. 1981.

Ketokivi, M., and S. Mantere. 2010. 'Two Strategies for Inductive Reasoning in Organization Research'. *Academy of Management Review,* 35: 3153.

Knorr Cetina, K. 1997. 'Sociality with Objects: Social Relations in Postsocial Knowledge Societies'. *Theory Culture and Society,* 14: 1.

Knorr Cetina, K. 1999. *Epistemic Cultures: How the Sciences Make Knowledge.* Cambridge, MA: Harvard University Press.

Krupp, S., and P. Schoemaker. 2014. *Winning the Long Game: How Strategic Leaders Shape the Future*. New York: Public Affairs.

Latour, B. 1987. *Science in Action*. Cambridge, MA: Harvard University Press.

Latour, B. 2004. *Politics of Nature*. Cambridge, MA: Harvard University Press.

Lave, J., and E. Wenger. 1991. *Situated Learning: Legitimate Peripheral Participation*. Cambridge: Cambridge University Press.

Leifer, R., C. McDermott, G. O'Connor, L. Peters, M. Rice, and W. Veryzer. 2000. *Radical Innovation: How Mature Companies Can Outsmart Upstarts*. Boston, MA: Harvard Business School Press.

Leonard-Barton, D. 1995. *Wellsprings of Knowledge: Building and Sustaining the Sources of Innvoation*. Boston, MA: Harvard Business School Press.

Leveson, N., N. Dulac, K. Marais, and J. Carroll. 2009. 'Moving beyond Normal Accidents and High Reliability Organizations: A Systems Approach to Safety in Complex Systems'. *Organization Studies*, 30: 227–49.

Lindbloom, C. 1959. 'The Science of Muddling Through'. *Public Administration Review*, 19: 79.

Locke, K., K. Golden-Biddle, and M. Feldman. 2008. 'Making Doubt Generative: Rethinking the Role of Doubt in the Research Process'. *Organization Science*, 19: 907.

Lynn, G., J. Morone, and A. Paulson. 1996. 'Marketing and Discontinuous Innovation: The Probe and Learn Process'. *California Management Review*, 38: 8–37.

McNamee, L., and F. Ledley. 2012. 'Patterns of Technological Innovation in Biotech'. *Nature Biotechnology*, 30: 937.

Magnani, L. 2001. *Abduction, Reason, and Science: Processes of Discovery and Explanation*. New York: Kluwer Academic/Plenum Publishers.

Malmberg, A., and D. Power. 2005. '(How) Do (Firms in) Clusters Create Knowledge'? *Industry and Innovation*, 12: 409.

Mayr, E. 2000. 'Biology in the Twenty-First Century'. *BioScience*, 50: 895.

Merriam-Webster's Dictionary of Synonyms. 1984. Springfield, MA: Merriam-Webster.

Meyer, M. and A. DeTore. 1999. 'Product Development for Services'. *Academy of Management Executive*, 13: 64–76.

Mumford, L. 1936. *Technics and Civilization*. London: Routledge & Kegan Paul.

Murray, F., and S. O'Mahony. 2007. 'Exploring the Foundations of Cumulative Innovation: Implications for Organization Science'. *Organization Science*, 18: 1006.

Nambisan, S., and M. Sawhney. 2011. 'Orchestration Processes in Network-Centric Innovation: Evidence from the Field'. *Academy of Management Perspectives* (Aug.): 40.

Nelson, R. 2005. *Technology Institutions and Economic Growth*. Cambridge, MA: Harvard University Press.

Nembhard, I., J. Alexander, T. Hoff, and R. Ramanujam. 2009. 'Why does the Quality of Health Care Continue to Lag? Insights from Management Research'. *Academy of Management Perspectives* (Feb.): 24.

Nesher, D. 2001. 'Peircian Epistemology of Learning and the Function of Abduction as the Logic of Discovery'. *Transactions of the Charles S. Peirce Society*, 37: 23.

Ng, R. 2004. *Drugs: From Discovery to Approval.* Hoboken, NJ: John Wiley.

Nightingale, P. 2004. 'Technological Capabilities, Invisible Infrastructure and the Un-social Construction of Predictability: The Overlooked Fixed Costs of Useful Research'. *Research Policy,* 33: 1259.

Orlikowski, W. 2002. 'Knowing in Practice: Enacting a Collective Capability in Distributed Organizing'. *Organizational Science,* 13: 249.

Orlikowski, W., and J. Yates. 2002. 'It's about Time: Temporal Structuring in Organizations'. *Organization Science,* 13: 684.

Orr, J. 1996. *Talking about Machines.* Ithaca, NY: ILR/Cornell University Press.

Owen-Smith, J., and W. Powell. 2004. 'Knowledge Networks as Channels and Conduits: The Effects of Spillovers in the Boston Biotechnology Community'. *Organization Science,* 15: 5.

Paavola, S., K. Hakkarainen, and M. Sintonen. 2006. 'Abduction with Dialogical and Trialogical Means'. *Logic Journal of IGPL,* 14: 137.

Patriotta, G. 2004. *Organizational Knowledge in the Making: How Firms Create, Use and Institutionalize Knowledge.* Oxford: Oxford University Press.

Pavitt, K. 1987. 'The Objectives of Technology Policy'. *Science and Public Policy,* 14: 182.

Pavitt, K. 1999. *Technology, Management and Systems of Innovation.* Cheltenham: Edward Elgar.

Penrose, E. 1959. *The Theory of the Growth of the Firm.* Oxford: Oxford University Press.

Pentland, B., T. Haerem, and D. Hillison. 2011. 'The (N)ever-Changing World: Stability and Change in Organizational Routines'. *Organization Science,* 22: 1369–83.

Perrow, C. 1986. *Complex Organizations,* 3rd edn. New York: Scott Foresman.

Piepenbrink, A. 2015. *Order without Authority: The Self-Organizing Innovation Ecology of a Standard Developing Organization.* Working Paper. Rennes: ESC Rennes.

Pisano, G. 2006. *Science Business: The Promise, the Reality, and the Future of Biotech.* Boston, MA: Harvard Business School Press.

Pisano, G. 2010. 'The Evolution of Science-Based Business: Innovating How we Innovate'. *Industrial and Corporate Change,* 19: 465.

Plowman, D., L. Baker, T. Beck, M. Kulkarni, S. Solansky, and D. Travis. 2007. 'Radical Change Accidentally: The Emergence and Amplification of Small Change'. *Academy of Management Journal,* 50: 515.

Powell, W., K. Koput, and L. Smith-Doerr. 1996. 'Inter-Organization Collaboration and the Locus of Innovation: Networks of Learning in Biotechnology'. *Administrative Science Quarterly,* 41: 116.

Prinz, F., T. Schlange, and K. Asadullah. 2011. 'Believe It or Not: How Much Can We Rely on Published Data on Potential Drug Targets'. *Nature Reviews: Drug Discovery,* 10: 712.

Rheinberger, D. 1992. 'Experiment, Difference, and Writing: Tracing Protein Synthesis'. *Studies in the History and Philosophy of Science,* 23: 305–13.

Rosenkopf, L., and A. Nerker. 2001. 'Beyond Local Search: Boundary-Spanning Exploration, and Impact in the Optical Disk Industry'. *Strategic Management Journal,* 22: 287.

Rothaermel, F., and M. Alexandre. 2009. 'Ambidexterity in Technology Sourcing: The Moderating Role of Absorptive Capacity'. *Organization Science*, 20: 759–80.

Rotman, D. 2014. 'Shining Light on Madness'. *MIT Technology Review*, 117 (July/ Aug.): 35.

Roussel, P., K. Saad, and T. Erickson. 1991. *Third Generation R&D: Managing the Link to Corporate Strategy*. Boston, MA: Harvard Business School Press.

Sammut, S. 2005. 'Biotechnology Business and Revenue Models: The Dynamics of Technological Evolution and Capital Markets Ingenuity', in Robert Burns (ed.), *The Business of Healthcare*, 190–222. Cambridge: Cambridge University Press.

Scannell, J., A. Blanckley, H. Boldon, and B. Warrington. 2012. 'Diagnosing the Decline in Pharmaceutical R&D Efficiency'. *Nature Reviews Drug Discovery* (Mar.): 191.

Schon, D. 1983. *The Reflective Practitioner: How Professionals Think in Action*. New York: Basic Books.

Simon, H. 1977. *Models of Discovery and Other Topics in the Methods of Science*. Dordrecht: D. Reidel Publishing.

Singer, E. 2009. 'Interpreting the Genome'. *Technology Review* (Jan./Feb.): 48–53.

Siren, C., M. Kohtamaki, and A. Kuckertz. 2012. 'Exploration and Exploration Strategies, Profit Performance, and the Mediating Role of Strategic Learning: Escaping the Exploitation Trap'. *Strategic Entrepreneurship Journal*, 6: 18.

Snowden, D., and M. Boone. 2007. 'A Leader's Framework for Decision Making'. *Harvard Business Review* (Nov.): 69.

Souder, W. 1987. *Managing New Product Innovations*. Lexington, MA: Lexington Press.

Stacey, R. 1995. 'The Science of Complexity: An Alternate Perspective for Strategic Change Processes'. *Strategic Management Journal*, 16: 477.

Star, S., and J. Griesemer. 1989. 'Institutional Ecology, "Translations", and Boundary Objects: Amateurs and Professionals in Berkeley's Museum of Vertebrate Zoology, 1907–39'. *Social Studies of Science*, 19: 387.

Stokes, D. 1997. *Pasteur's Quadrant: Basic Science and Technological Innovation*. Washington, DC: Brookings.

Su, Y. 2013. *Transforming Academic Knowledge for Drug Innovation: A Practice-Based View of Objects, Entrepreneurs, and Institutions*. Dissertation. Rutgers University, Newark, NJ.

Su, Y., and D. Dougherty. 2015. *The Multiple Dynamics of Divergence and Convergence between Basic and Applied Science for Innovation*. Working Paper. Singapore: Singapore Management University.

Sull, D. 2001. 'From Community of Innovation to Community of Inertia: The Rise and Fall of the U.S. Tire Industry', *Academy of Management Proceedings*, BPS, L1.

Taylor, J., and E. Van Every. 2000. *The Emergent Organization: Communication as its Site and Surface*. Mahwah, NJ: Erlbaum.

Tidd, J., and J. Bessant. 2009. *Managing Innovation: Integrating Technological, Market and Organizational Change*, 4th edn. Chichester: John Wiley.

Tsoukas, H. 2005. 'Noisy Organizations: Uncertainty, Complexity, Narrativity', in Haridimos Tsoukas (ed.), *Complex Knowledge: Studies in Organizational Epistemology*, 338–77. Oxford: Oxford University Press.

Tsoukas, H., and K. Dooley. 2011. 'Introduction to the Special Issue: Towards an Ecological Style: Embracing Complexity in Organizational Research'. *Organization Studies*, 32: 729.

Tsoukas, H., and C. Knudsen. 2005. 'The Conduct of Strategy Research: Meta-Theoretical Issues', in Haridimos Tsoukas (ed.), *Complex Knowledge: Studies in Organizational Epistemology*, 338–77. Oxford: Oxford University Press.

Turro, N. 1986. 'Geometric and Topological Thinking in Organic Chemistry'. *Agnew Chemistry, International Edition English*, 25: 882–901.

Tushman, M., and C. O'Reilly. 1997. *Winning through Product Innovation*. Boston, MA: Harvard Business School Press.

Tushman, M., and L. Rosenkopf. 1992. 'On the Organizational Determinants of Technological Change: Toward a Sociology of Technological Evolution', in Barry Staw and Larry Cummings (eds), *Research in Organization Behavior*, xiv. 311–47. Greenwich, CT: JAI Press.

Van de Ven, A. 1986. 'Central Problems in the Management of Innovation'. *Management Sciences*, 32: 590–607.

Van de Ven, A., D. Polley, R. Garud, and S. Venkataraman. 1999. *The Innovation Journey*. New York: Oxford University Press.

Vickers, G. 1968. *Value Systems and Social Process*. New York: Basic Books.

von Hippel, E. 1988. *The Sources of Innovation*. New York: Oxford University Press.

Wareham, J., P. Fox, and J. Cano Giner. 2014. 'Technology Ecosystem Governance'. *Organization Science*, 25: 1195.

Webster's New World College Dictionary. 2007. Cleveland, OH. Wiley.

Weick, K. 1979. *The Social Psychology of Organizing*, 2nd edn. Reading, MA: Addison-Wesley.

Weick, K. 1995. *Sensemaking in Organizations*. Thousand Oaks, CA: Sage.

Weick, K. 2005. Organizing and Failures of Imagination. *International Public Management Journal*, 8: 425–38.

Weick, K., and K. Roberts. 1993. 'Collective Mind in Organizations: Heedful Interrelating on Flight Decks'. *Administrative Science Quarterly*, 38: 357.

West, W., and P. Nightingale. 2009. 'Organizing for Innovation: Towards Successful Translational Research'. *Trends in Biotechnology*, 27: 558.

Yaqub, O., and P. Nightingale. 2012. 'Vaccine Innovation, Translational Research, and the Management of Knowledge Accumulation'. *Social Science and Medicine*, 75: 2143.

Index

Tables are indicated by an italic *t* following the page number.